Willis Stanley Blatchley

Gleanings from Nature

Willis Stanley Blatchley

Gleanings from Nature

ISBN/EAN: 9783337026059

Printed in Europe, USA, Canada, Australia, Japan

Cover: Foto ©berggeist007 / pixelio.de

More available books at **www.hansebooks.com**

THE BROOKSIDE.

Stream from Shawnee Cave, 100 yards below mouth of Cave. See the poem
"A Rest by the Brookside;" also page 121.

GLEANINGS FROM NATURE

BY

W. S. BLATCHLEY

"I make it my business to extract from Nature whatever nutriment she can furnish me, though at the risk of endless iteration. I milk the earth and the sky . . . . . I sift the sunbeams for the public good."
—*Thoreau.*

INDIANAPOLIS
The Nature Publishing Company
1899

"Nature never did betray
The heart that loved her; 'tis her privilege,
Through all the years of this our life, to lead
From joy to joy; for she can so inform
The mind that is within us, so impress
With quietness and beauty, and so feed
With lofty thoughts, that neither evil tongues,
Rash judgments, nor the sneers of selfish men,
Nor greetings where no kindness is, nor all
The dreary intercourse of daily life
Shall e'er prevail against us, or disturb
Our cheerful faith that all which we behold
Is full of blessing."

— *Wordsworth.*

"Whether it be the crested tit defying the chilliest blast of January; violets mantling the meadow banks in April; thrushes singing their farewell summer songs, or dull and dreary, dim December days it matters not—they never repeat themselves, or else I am daily a new creature. Nor sight nor sound but has the freshness of novelty, and one rambler, at least, in his maturer years is still a boy at heart."—*C. C. Abbott.*

To the 800,000 boys and girls on the farms of Indiana this little volume is inscribed, with the hope that it may create in some of them an interest in the many objects of Nature which surround them and so cause them to be less

"Blind to her beauties everywhere revealed,"

less prone to

"Tread the May-flower with regardless feet."

PREFACE.

This volume deals with a few of the many natural objects which are found in all parts of Indiana. It is based upon studies made in the fields and woods of the Hoosier State during the past ten years. The aim has been to present, in language which all can understand, facts concerning some of the more common plants and animals which are our friends, our helpers and our neighbors—and which, like ourselves, are but a part and parcel of the Universe The only technical terms used are the scientific names of some of the objects mentioned. These are printed in Italics and can be readily passed over by all to whom they are unintelligible.

The contents have, for the most part, appeared elsewhere as isolated articles, notably in the Terre Haute Gazette, the Indianapolis Sunday Journal, The Indiana Farmer and the Popular Science Monthly. For the present occasion they have, for the first time, been brought together, and have been carefully revised and enlarged.

The volume is sent forth with the hope that among the farmers of the future and the teachers of country schools it will, at least, have a welcome; for the author knows by experience, both on the farm and in the school room, that the possession of a better knowledge of nature by country youths is one of the crying needs

of the hour. With such a knowledge generally dif-
fused there would be less dissatisfaction with country
life and fewer farmers' sons and daughters would flock
to the cities, because, as a recent writer expresses it,
"they wish to get rid of the prosy, stunting, isolated
life on the farm." With a knowledge of some of
nature's objects and a desire to ferret out for themselves some of her secrets, they would have something of which to talk and think besides crops, stock,
work, neighborhood gossip and local politics, and the
attractions of the city would seldom excel those to be
found on the old homestead.

CONTENTS.

A REST BY THE BROOKSIDE.

Dreaming dreams of other days,
  Thinking thoughts of long ago,
List'ning to the robin's lays
  And the cawing of the crow.

Ivy flowers beside me peep
  Upward through the ether blue,
Seeing stars which ever keep
  Hidden close from human view.

Bumble-bees around me drone,
  Butterflies beside me flit;
From the woods in cheery tone
  Comes the call of crested tit.

Swallows swiftly cleave the air
  Chasing insects on the wing;
Scolding chats with saucy dare
  Make the copse and welkin ring.

Gurgling waters at my feet
  Quickly o'er their pebbly bed
Leap and plunge; and onward meet
  Other streams by springlets fed.

Odors sweet on every breeze
  Come to me from wild-wood flowers,
While the blossoms from the trees
  Fall around in fragrant showers.

Happy moments thus I gaze
  Heavenward; and younger grow,
Dreaming dreams of other days,
  Thinking thoughts of long ago.

HARBINGERS OF SPRING.

When the month hand on the dial plate of the year
points to March who, in the latitude of Indiana, is not
daily expecting spring? What human being is not
made glad when it finally arrives? Four months of
biting winds and hoar frosts; months in which the
skies are almost daily overcast with dull dreary clouds;
months of alternate rains and sleets and snows, are
enough to cause an intense longing for change in the
human mind and to bring to it a glow of happiness
when the first warm breezes blow up from the gulf
and man can say with reason—"Spring has come
again." Then the dormant energies within us spring
into new life. The doors and windows of our houses
are thrown open wide. Smiles are seen on faces to
which for the most part they are strangers. Wee tots
of children run unattended up and down the streets
and laugh and shout with joy. Matrons forget or
cast aside set social rules and stop and chat in one
another's dooryards. Fancied class distinctions, based
on wealth or "blue blood," are forgotten and, for the
time being, the members of the human family are
more akin than at any other season of the year. All
are enjoying a common blessing, for spring comes

alike to rich and poor, to high and low, and all can revel in its presence.

To one accustomed to visit the woods and fields during March there appear many unerring signs of the coming spring-time which, to persons living in towns and cities, are often unnoticed and unknown. The growth and flowering of certain wild plants; the awakening from their winter's sleep of reptiles, frogs and insects; the arrival of the first migrating birds, are to the careful observer sure harbingers of the close approach of the vernal season. If in March there occurs, as often happens, several successive days of warm weather more than a dozen kinds of wild plants will come into bloom. They are the fore-runners or vanguard of the eight to nine hundred species of flowering plants which, in any county of Indiana, open their petals in successive rotation between March the first and mid-October.

Perhaps the earliest flowers of spring are those of the red or swamp maple, *Acer rubrum* Linn., a medium-sized tree which grows in abundance in damp lowland soil. This maple is often brought into the cities and palmed off on unsuspecting buyers of shade trees as the soft or white maple, *Acer saccharinum* Linn. Both of these trees differ in their habits of flowering from the rock or sugar maple, *Acer saccharum* Marsh, in that their blossoms appear before their leaves. The flowers of the red maple are a handsome deep red in color and are arranged on very short stems in little clusters near the ends of the branches. They sometimes open in February, as they are formed in autumn, and, protected only by the enveloping bud scales, are

II

SNOW TRILLIUM.
Trillium nirale Riddell.

ready to unfold as soon as the dormant sap of the parent tree is awakened by a genial south wind. The blossoms of the soft maple usually open a fortnight later than those of the red maple. They are yellowish-green in color and are borne on longer stems.

Of the herbs which bloom in March there are two which, in central Indiana, vie with the red maple in producing the first wild flowers of spring. They are the little snow trillium, *Trillium nivale* Riddell, and a species of *Draba*, or whitlow-grass, both inconspicuous plants and known for the most part only to botanists and close observers of nature. The snow trillium belongs to the Lily family and grows only at the base of rocky cliffs or in crevices along the sides of ravines which have a sunny southern exposure. It seldom exceeds four inches in height, and, as its name, "trillium," indicates, has its parts in threes or multiples of three. Three dark green, ovate leaves grow in a whorl at the summit of the slender stem, and from their midst springs the stalk of the solitary flower. This is composed of three narrow green sepals, three oblong pure white petals, each about an inch in length, six yellow stamens, three styles and an ovary or seed pod containing three cells, each with a number of minute seeds. The plant springs from the ground and blooms in less than forty-eight hours, and where one day all is brown and sere, on the second day thereafter may be found an abundance of these little trilliums—true earth-born harbingers of the approaching springtime.

They may be found in blossom as early as March 2d, and are often in their prime by the tenth of

the month. Hovering above them on both dates
have been seen specimens of *Vanessa antiopa* Linn.,
that handsome velvety-brown butterfly, called the
mourning cloak, or Camberwell beauty. This is the
most common of the five species of butterflies which
in Indiana pass the cold season in the perfect or winged
stage. For four long months they remain securely
hidden in crevices of rocks or logs. When called

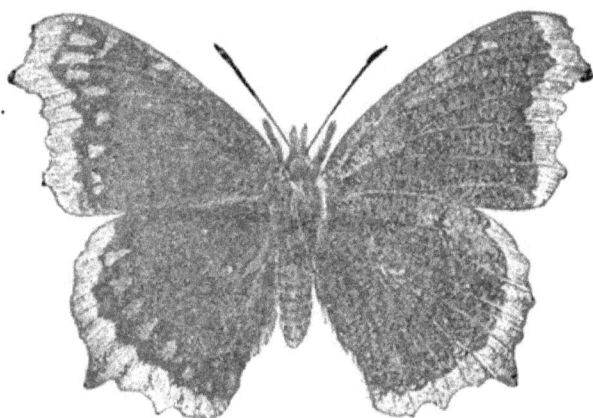

Fig. 1—Camberwell Beauty. (After Harris.)
(The under side of wing is shown on the right.)

forth by the warm spring breezes they find ready and
waiting for them the snow trilliums with their store
of honey, and, after their prolonged fast, they no doubt
feast bountifully thereon. Thus is seen an example
of that mutual interdependence existing everywhere
among the various objects of nature, for never does an
insect come forth until its food plant is ready; and,
on the other hand, the plant seldom blooms but an
insect appears ready to aid in its fertilization.

Springing as it does from crevices and crannies in rocky cliffs, reaching maturity and flowering as it does when all nature is destitute of bloom, the first snow trillium of spring carries ever with it a reminder of those lines of Tennyson:

> "Flower in the crannied wall,
>   I pluck you out of the crannies;
>   Hold you here, root and all, in my hand,
>   Little flower, but if I could understand
>   What you are, root and all, and all in all,
>   I should know what God and man is."

Two species of Draba, or whitlow-grass, grow in central Indiana, viz., *D. caroliniana* Walt. and *D. verna* Linn. They are the smallest members of the *Cruciferæ*, or Mustard family, and have their minute and hairy oblong leaves clustered in a rosette close to the ground. From the center of this rosette the leafless flower-stalk springs. The flowers are white, ten to fifteen in number, and have the parts in fours, except the stamens, which are six. The plants are found on dry, sandy hillsides in open fields. Rising less than three inches above the ground, they bloom on the first warm days of March, and their seeds are ripened by mid-April. Their work is, therefore, over before that of many plants is begun. They succeed in the struggle for existence by being first upon the scene of action. Drinking long and deep of the bright spring sunshine, they soon give way to their competitors, but not before their life's duty—the perpetuation of their kind—has been fulfilled.

On soft, springy banks one also finds in earliest spring the curiously formed flower of the skunk cab-

bage, *Symplocarpus fœtidus* Nutt. It is, as it were,
a leafless flower, barely rising out of the ground. At
first the only semblance of a leaf is the enveloping
spathe (like that of a calla lily), which is variously
striped and spotted with purple and yellowish green,
and has its top incurved or bent over like a bird's
beak to protect the enclosed flowers from any wintry
blast which may arise. The true flowers are numerous, small and inconspicuous, and are borne on a
thick, fleshy spadix or central axis within the spathe.
The plant, after flowering, sends up a thick cluster
of large ovate leaves, eighteen inches or more long,
which surround the spathe. These leaves, when
bruised, give off an odor resembling somewhat both
that of a skunk and a mess of strong onions, or a
kind of potpourri of the two, hence the common name
of "skunk cabbage" by which the plant is known.
Rich in pollen, the flowers of this plant furnish the
first meal of spring to many a honey-bee; for the first
warm day of March tempts these busy insects forth
in numbers, and they find their way unerringly to
the few flowers then in bloom.

Other plants there are, more striking and more
beautiful than those mentioned, which soon open wide
their petals to the sunshine of spring. Among the
more common of them is the hepatica, or liverwort,
Hepatica triloba Chaix.; the turkey pea, or "pepper
and salt," *Erigænia bulbosa* Michx.; the spring beauty,
Claytonia virginica Linn., and the cowslip, or marsh
marigold. *Caltha palustris* Linn., but they are followers,
not leaders in the floral army. When they blossom
spring is here. But the flowers of the red maple,

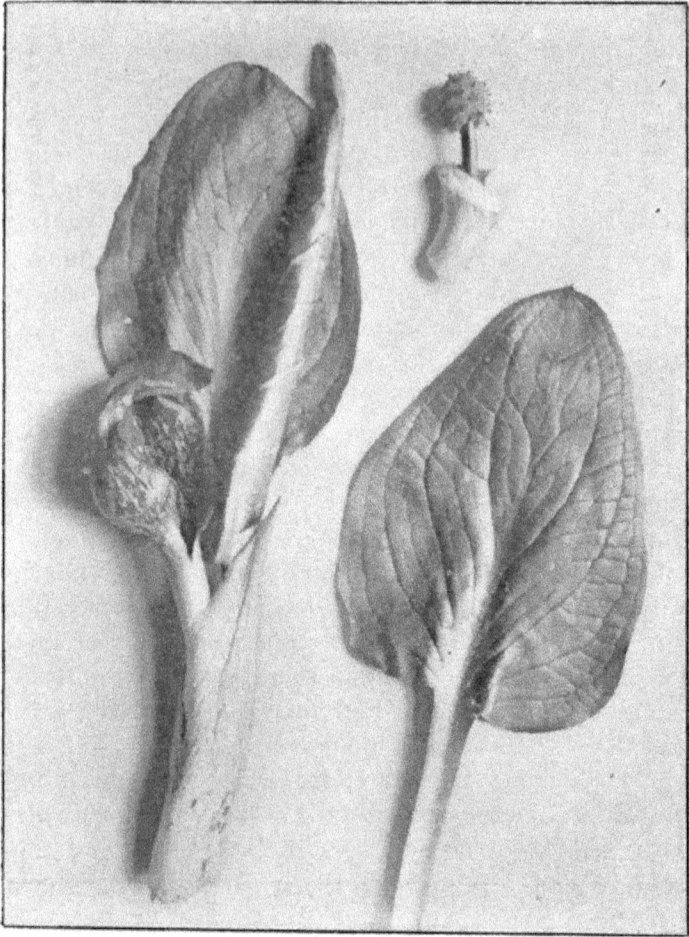

III

SKUNK CABBAGE.

Symplocarpus fœtidus Nutt.

snow trillium, whitlow-grass and skunk cabbage are
the pioneers, the true harbingers, which herald its
approach.

A great awakening also takes place among the
varied forms of animal.life on the first warm days of
March.   Among insects the wherrymen, those long-
legged water bugs which go skipping so easily and
rapidly over the surface of quiet pools, and the whirling
beetles which in vast colonies go circling round and
round on the water, are the first ones out.   Fuzzy gnats,

> "Old back-bent fellows,
>   In frugal frieze coat drest,"

come forth from their snug retreats beneath the bark
of the beech and other logs, and, swarming in the air,
carry on a sort of rhythmical courtship, flitting up
and down in the same vertical plane in a dreamy,
dancing sort of motion.

Beetles, of which, in any county in Indiana, fully
three hundred kinds survive the cold season in the
winged stage, crawl out from their winter hiding
places and a-wooing go, buzzing and humming with
extra energy to attract the notice of others of their
kind.   With the hibernating butterflies mentioned
above, and numerous kinds of wild bees and flies,
they frequent the freshly cut stumps of the sugar
maple, where they sip eagerly the sweet exuding sap.

Higher in the scale of animal life omens of approaching spring may be seen in the movements of
fishes.   Thrilled with the impulse of migration, many
of the smaller species begin in February and March
to ascend small streams and brooks, where, beneath

the shelving banks and in the still waters of the
deeper pools, they make their summer homes.

Frogs are among the best of weather prophets.
They seem to know intuitively when the spring is
full on its way. Long before the frost is wholly out
of the ground the sluggish blood within their veins
begins to tingle, and they greet the first dawn of
spring with a mighty chorus—a blare of welcoming
trumpets, as it were—in which the bull-frog furnishes
the bass and the little cricket frog or "peeper," *Acris
gryllus crepitans* Baird, the shrill whistling tenor. No
sound of nature so loudly or so surely proclaims the
advent of spring as this full symphony of frog music
heard from some woodland pond.

The arrival of the first migrant birds is also a sure
symbol of the coming spring. About 180 species pass
northward through Indiana between February 15 and
the 10th of May. In addition to these at least seventy-five kinds stop in the State and nest and rear their
young. Wild geese and ducks are the first ones to
be seen northward bound. Impelled by the pairing
instincts, thousands of squads of these water birds
start in February from the sunny lakes and lagoons
of the South for the still cold and cheerless breeding
grounds that extend from the Northern States through
British America to the Arctic seas. The wild geese
fly, as is well known, in a V-shaped line, with the
apex forward. Their leader is a strong-winged gander, who keeps his place at the point of the V, and
the clarion-toned "honk" with which he gives his
orders is the first note of that coming bird chorus,
which, starting from the gulf, will, with the south

winds, soon sweep northward through field and forest
in an unbroken wave to the very pole itself.

In the close wake of the larger water fowl come
the snipe and the woodcock, *Philohela minor* (Gmel.)
—the latter arriving so early that a full set of its eggs
was once found by the writer on the 28th of March.

Fig. 2—The Flicker or Yellow-hammer. (After Beal.)

Among the land birds two of the first to arrive are
the flicker, or yellow-hammer, *Colaptes auratus* (Linn.),
and the red-shouldered blackbird, *Agelaius phœniceus*
(Linn.); and the prolonged "wick-a-wick-a-wick" of
the former and the clear, ringing "puck-e-e-eet" of

2

the latter are familiar greetings given the rambler in open woods in early March. These birds are soon followed by the handsome but little-known fox sparrow, *Passerella iliaca* (Merr.), whose rich strains, heard from the underbrush along the streams, form the first real song of spring.

Some years winter lingers unusually long in the lap of spring, and two-thirds of March may come and go and but few of the harbingers above mentioned be seen. But that day of great awakening, which in the temperate zone comes each year to all animate things, in time arrives. We should expect it, should rejoice to see it, should give it hearty greeting. In the words of Thoreau: "Measure your health by your sympathy with morning and spring. If there is no response in you to the awakening of nature; if the prospect of an early morning walk does not banish sleep; if the warble of the first bluebird does not thrill you, know that the morning and spring of your life are past. Thus may you feel your pulse."

TWO FOPS AMONG THE FISHES.

I.—THE RAINBOW DARTER.

"Little fishy in the brook."

Not the one "daddy caught with a hook," but another, too small for the hook, too small for the frying-pan, too small for aught else but beauty, and gracefulness of form; and yet not the young of a larger fish, but full grown of himself. In every brook in the State he may be found, yea, even in the rill, no more than a foot in width, which leads away from the old spring-house on the hillside. You will not find him swimming about like the minnows in the still deep water of the stream, but where the clear cold water is rushing rapidly over the stones of a ripple he makes his home. There he rests quietly on the bottom, waiting patiently for his food, the larvæ or young of gnats, mosquitoes and other such insects, to float by.

If you attempt to catch him, or your shadow suddenly frightens him, with a sweep of his broad pectoral or breast fins, he moves quicker than a flash a few feet farther up the stream and then as suddenly comes to a stop, and resumes his quiet "thoughtful" attitude. If you persist in your attempt to capture him he will dart under a small stone or submerged leaf, where, like the foolish ostrich which when pursued hides her head under her wing, no longer seeing you, he thinks himself secure.

On account of the shape of his body as well as on account of his rapid movements he has received the surname "darter." Belonging to the group which bear this surname, there are, in the eastern half of the United States, about 47 species or kinds, the largest of which, when full grown, measures only about six inches in length, while the smallest species never reaches a length of more than an inch and a half. They all have the same habits, and at least 29 kinds of them are found in Indiana; but the one of which I am writing, *Etheostoma cœruleum* Storer, is much the

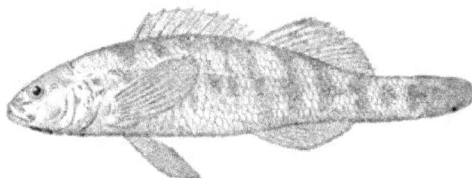

Fig. 3—Rainbow Darter.

more common. He is from two to two and a half inches in length, and, like the other members of his family, has two fins on his back; "dorsal" fins they are called by naturalists, the front one of which contains 10 short spines. During eight months of the year, the males and females dress alike in a suit of brownish olive which is striped on the sides with 10 or 12 narrow, black cross-bars, and more or less blotched on the back with darker spots. But on the first warm days of spring when the breezes blow up from the gulf, awakening the gypsy in our blood, the little male fish feels, too, their influence, and in him there

arises an irresistible desire to "a courting go." Like
most other beings of his sex he thinks his every-day
suit too plain for the important business before him.
It will, in his opinion, ne'er catch the eye of his lady
love. So he dons one of gaudy colors and from it
takes his name—the rainbow darter—for in it he is
best known, as it not only attracts the attention of his
chosen one, but often also that of the wandering naturalist who happens along the stream.

The blackish bars of other seasons are changed to
indigo blue, while the space between them assumes a
hue of the brightest orange. The fins are broadly
edged with blue and have the bases orange, or orange
and scarlet, while the cheeks assume the blue and the
breast becomes an orange. Clad in this suit he ventures forth on his mission, and if successful, as he
almost always is, the two construct a nest of tiny
stones in which the eggs of the mother fish are laid
and watched over with jealous care by both parents
until in time there issue forth sons destined some day
to wear a coat of many colors, and "darters" to be
attracted by those coats, as was their mother by the
one their father wore.

Although so abundant and so brilliant in the spring-
time, the rainbow darter is known to few but naturalists. The fishes in which the average country boy is
interested, are the larger ones—such as the goggle-
eye, the sucker, chub and sunfish—those which, when
caught, will fill up the string and tickle the palate.

But there are, let us hope, among our farmers' sons
and daughters, some who are learning to take an
interest in the objects of nature which are beautiful,

as well as in those which are useful. To them I will
say, if you wish to see something really pretty, make
a seine from an old coffee sack or a piece of mosquito
netting, and any day in spring drag two or three ripples of the branch which flows through the wood's
pasture, and ten chances to one you will get some
"rainbows." By placing them in a fruit jar three-
fourths full of clear, cold water, and renewing the
water every few hours, they can be kept for several
days; but they can not bear the confinement long,
accustomed as they are to the free running stream
from which they were taken.

By taking the rainbow as the type of the darter
and studying closely its habits, both in captivity and
in the streams, much can be learned about a group
which, in the words of Dr. S. A. Forbes, "are the
mountaineers among fishes. Forced from the popu-
lous and fertile valleys of the river beds and lake bottoms, they have taken refuge from their enemies in
the rocky highlands where the free waters play in
ceaseless torrents, and there they have wrested from
stubborn nature a meager living. Although diminished in size by their continual struggle with the
elements, they have developed an activity and hardi-
hood, a vigor of life and a glow of high color almost
unknown among the easier livers of the lower lands."

II.—THE LONG-EARED SUNFISH.

Among the most brightly colored of all the fresh
water members of the finny tribe is the long-eared
sunfish, *Lepomis megalotis* (Raf.). When full grown

its length is about eight inches and the breadth one-half as much. The color is then a brilliant blue and orange, the former predominating above; the orange on the sides in spots, the blue in wavy, vertical streaks. The cheeks are orange with bright blue stripes; the fins with the membranes orange and the rays blue. Extending back from the hind margin of each cheek is a conspicuous blackish membrane termed an "ear-flap," which in this species is longer than in any other

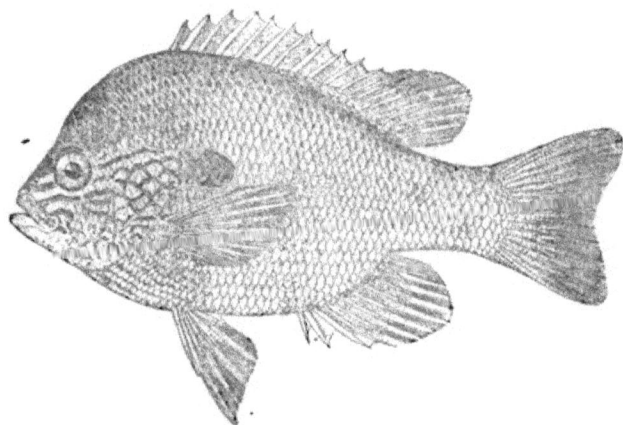

Fig. 4—Long-eared Sunfish.
(One-half natural size.)

of the sunfish family, whence the specific name, *megalotis*, from two Greek words meaning "great" and "ear".

Within the placid pools of the brooks and larger streams of the State this sunfish has its favorite haunts. Mid-summer is the time when its habits can be best observed. On a recent August morn I sat for an hour or longer on the banks of a stream, which flows through a wooded blue-grass pasture, and watched the denizens of its waters. A peaceful calm existed, the water being without a ripple and with scarce the semblance of a flow—the air without the shadow of a breeze. Dragon flies lazily winged their way across the pool, now resting daintily upon a blade of sedge or swamp grass, now dipping the tips of their abdomens beneath the surface of the water while depositing their eggs. The only sounds of nature were the buzz of a bumble-bee feeding among the flowers of the *Brunella* at my side, and an occasional drawl of a dog-day locust from the branches of the sycamore which threw a grateful shade about me.

The sunfish "hung motionless" in the water, their heads towards me, holding their position only by a slow flapping of their dorsal and pectoral fins. Their nesting time over, their season's labor ended, it was with them, as with many other beings, a time of languor.

These long-eared fishes are the lords and ladies of the respective pools wherein they abide. When they move other smaller fry clear the way. If a worm or gnat, falling upon the surface, tempts them, it is theirs. A leaf falls near them and is seemingly unnoticed—a fly, and how quickly their dormant energy is put into motion. With a dart and a gulp the insect is swallowed and a new stage of waiting expectancy is ushered in.

How admirably fitted their form for cleaving the water! They often seem to glide rather than propel themselves through its depths. Again, how swiftly the caudal fin moves when with straight unerring motion they dart upon their prey. At times one turns his body sideways and, with a slow, upward-gliding motion, moves toward some object on the surface which is doubtfully "good to eat." He even takes it into his mouth and then, not having faith in his power to properly digest it, ejects it with force, and turning quickly darts back to the friendly shadow of a bowlder beneath whose sides he has, in time of threatened danger, a safe retreat.

I throw a grasshopper into the pool. Like a flash six of the sunfish are after it. One reaches it a tenth of a second in advance of the others, and with a lightning-like gulp, which disturbs the serenity of the surface of the pool, swallows the kicking prey. The energy of the sun's heat and light, stored in grass, transmitted to move muscles in gigantic leaps, will, in a short time, wag a caudal fin and propel the owner through these watery depths.

Years are thus doubtless spent by these long-eared sunfish in a dreamy sort of existence, their energies quickened by the vernal season and growing duller on the approach of winter. Excepting the times when they are tempted by a wriggling worm on some boy's hook, theirs is a life exempt from danger. A kingfisher glancing down from his perch on the bent sycamore limb may, at times, discern them and lessen their ranks; but, methinks, the chub minnows, with fewer spines in their dorsal fins, are more agreeable to the kingfisher's palate. With all the tints of the rainbow gleaming from their sides they move to and fro, the brilliant rulers of these quiet pools.

The king or monarch of those noted was most

gorgeously arrayed. In addition to the hues above . described, a streak of emerald bordered his dorsal and caudal fins and was bent around the edge of his upper lip—a green mustache, as it were. By tolling them with occasional bits of food I drew him and his retinue close into shore. There, for some time they rested, watching eagerly for additional morsels. As I was leaving I plucked from my sleeve an ant and threw it towards them. A dart, a gurgle, a gulp—the leader had leaped half his length from the water, and the ant was forever gone. The ripples receded and finally disappeared, and the last scene in this tragedy of nature was at an end.

SNAKES.

I.—SNAKES IN GENERAL.

Snake season will soon be here once more. Even now* the editors of the country newspapers are poring over the musty pages of some ancient natural history or seeking amidst those still more musty convolutions of their brains in which their natural history facts are stored, to find a basis for one of the annual snake lies, which, like the dandelions, are sure to appear when spring approacheth. For next to "fish stories," newspaper "snake lies" are sure to be appreciated and believed in by a certain class of readers. True it is that the editor or reporter does not always clothe the lie in all its after habiliments, but each reader on repeating it to his neighbor adds a garment, until out of "whole cloth," as it were, the lie becomes a finished product and is repeated so often that it is finally believed as gospel truth.

Examples of such snake lies are the "hoop snake" which, taking its tail in its mouth, rolls rapidly on-

Some Common Snake Lies.

ward until it strikes a tree into which it darts its tail, poisoning the sap and causing the death of the tree; the "glass snake" which, when approached, breaks into a score of pieces, and when unobserved quickly joins

---

*Terre Haute Gazette, April 2d, 1892.

them again in the proper order and moves rapidly
away; the black snake "eight feet long and as thick
as your wrist"; the spreading viper "the most poisonous of all snakes," and of which one writer has said:
"When approached it becomes flat, appears of different colors, and opens its mouth hissing. Great care
must be taken not to enter the atmosphere which surrounds it. It decomposes the air, which, imprudently
inhaled, produces languor, the person wastes away,
the lungs are affected and in the course of four months
he dies of consumption." That this last story, or
something akin to it, is commonly believed, is proven
by the fact that a prominent citizen once told the
writer that the breath of the spreading viper had
caused him a two weeks' spell of sickness.

Many other "snake lies" the writer has heard, and,
to tell the truth, believed in, until he came to get his
knowledge first hand by studying the creatures in
their chosen haunts, when he saw how unworthy of
belief many of these stories are. For example, taking
the four above noted, and tracing each back to its
source, we find that the common house or milk snake,
Ophibolus doliatus triangulus (Boie), while crawling,
occasionally raises the middle of its body above the
ground, as does the measuring or loop worm, and this
fact gave rise to the story of the "hoop snake."

The "glass snake" is a lizard, *Ophisaurus ventralis*
(L.), which, like other reptiles of that class, chooses at
times when captured by the tail, to drop that portion
of the body rather than remain a captive; but, as to
coupling it on again, no person with any regard for
the truth will swear he ever saw it done.

No black snake over six feet and a few inches long has as yet been recorded in any scientific work. Twice within the past five years black snakes, kept in captivity by the writer, have escaped, and were killed shortly afterwards near the center of the city. On both occasions the daily papers noted the killing but in each instance, if we take the newspaper measurement as correct, the snake had increased in length nearly a foot and a half during the twenty-four hours intervening between its escape and death.

As for the spreading viper, although it is true that it flattens its body and hisses when approached, yet its bite is perfectly harmless as it is destitute of poison fangs; and its colors are as unchangeable and its breath as unproductive of disease as are those of the leading gander of the barnyard flock which hisses when we approach his domain.

The best way to show the falsity of many of the beliefs concerning the harmfulness of snakes is to record a few facts concerning the life history and habits of some of the more common species inhabiting Indiana. To begin, we will say that the usual belief that all snakes are hatched from eggs is an erroneous one. Many species, examples of which are the copperhead, rattle-snake and three or four kinds of garter-snakes, bring forth their young alive. The young of snakes, except in size and sometimes in color, resemble their parents and do not undergo a change or metamorphosis, as do the tadpoles, or young of frogs and salamanders. Those snakes which lay eggs deposit them in the earth, sand, or the humus of rotten logs, where they are left to be hatched by the moist heat of their surroundings. These eggs are yellowish white in color, and vary in size from $\frac{1}{4}$x$\frac{3}{4}$ inches

The Young of Snakes. up to that of a pigeon's egg. They are usually elliptical in form and have a tough leathery skin. The number varies with the different species, some laying as many as twenty at a time. The mother snake sometimes remains in the vicinity of the eggs until they are hatched. The young then accompany the mother for a time, and of certain species, it has been affirmed, that in time of danger the young escape down the throat of the mother. Of most snakes, however, the young, when hatched or born, are left to shift for themselves, and possibly not more than one in a hundred lives to be a year old, as they have many enemies among the other animals—even among their own kin.

If a snake be carefully examined many interesting facts concerning the structure of its body may be noted. Their long, slender, limbless forms are, on the upper side, covered with scales which overlap one another like the shingles on a roof. On the under side these scales are much larger and form a series of broad, overlapping plates which extend the full length of the body. These are technically known as ventral plates or scutes.

Many a person has, perhaps, wondered how an animal without limbs, wings or fins can move so rapidly and gracefully as does a snake. **The Structure of a Snake.** By examining carefully a snake's skeleton and noting its relation to these ventral plates one can easily understand how the movement is made. The skeleton consists merely of the skull, spinal column and ribs.

A pair of ribs extend downward from each vertebra of the spinal column, and to the lower ends of these ribs a ventral plate is attached by muscles. The snake then moves its ribs much as a millipede or "thousand-legged worm" moves its legs. The edges of a few of the ventral plates catch against any roughness on the surface over which the snake is crawling, and hold that part of the body while another part advances. Put a snake onto a smooth surface as ice or a polished floor and it will move with much difficulty, if at all. Hence, we see, that a snake in reality walks with its ribs.

The skin of a snake, scales, plates and everything, is shed several times a year. The first moult of the season usually takes place in the early spring, soon after the snake regains activity;

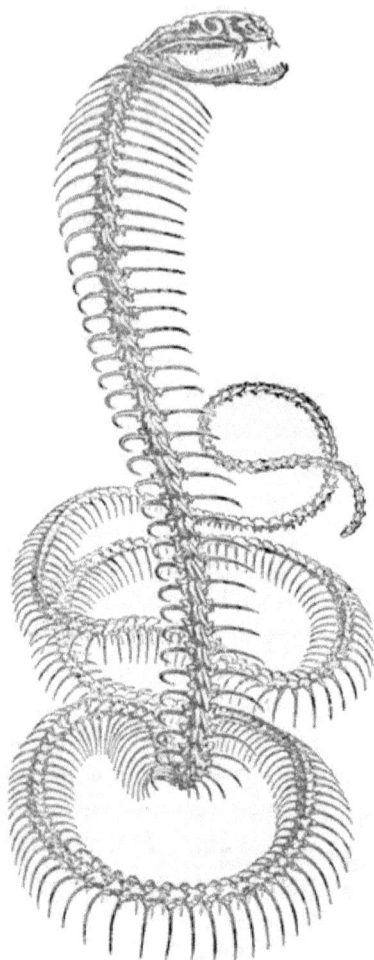

Fig. 5.— Skeleton of a Snake.

a second occurs in June or July, and often a third
in late summer or early autumn.   One can always
tell when the moulting is about to take place by the
color of the snake becoming very dull and an ap-
parent whitish film appearing over the eyes   In fact
the snake seems to be going blind.   The skin is shed
as a whole, a rent appearing on the back, and first
one end of the body and then the other being pulled
out of the old garment.   Even the cornea, or outer
surface of the eye, and the skin of the lips are shed.
The new skin is very bright and showy and the snake
is evidently proud of it, appearing much more lively
after moulting than before.   New skins are constantly being formed beneath old ones and the reptile
must keep one or two on hands for an emergency, as
the writer has by dissection found three skins on one
snake.

Many persons are needlessly frightened when a snake
darts out its tongue at them.   The tongue is nothing
but a thread-like muscle forked or divided for about
one-third of its length.   It lies on the middle of the
lower jaw and when at rest is covered with a sheath-
like membrane.   Soft and elastic in structure, it is
capable of being darted back and forth very rapidly.
Although the tongue is perfectly harmless the snake,
during past generations, seems to have learned that
man and some other animals are afraid of it, and so,
when irritated or molested, darts it in and out as a
means of defense.

Snakes have no outer ears and no eyelids.   To the
latter fact is due the "cold stony glare of the serpent"
Their jaw bones or mandibles are held together only

by ligaments. This enables them to open the mouth
very widely and to swallow animals much larger in
diameter than themselves. They swallow all food
whole without mastication. In this they are aided
by a copious flow of saliva which lubricates their
prey, and causes it to pass into the stomach more
readily. Whenever they catch a frog or other animal
with limbs they manipulate it in such a manner as to
enable them to swallow it head first. The limbs of
the victim are thus pressed close to its body and the
act of swallowing is but little hindered by their
presence.

Snakes are "cold blooded" animals; i. e., their
bodily temperature is not constant like that of man,
but varies with the temperature of the air which they
breathe. On that account they become sluggish in
late autumn, and, seeking a crevice in a rock or hole
in the ground, they crawl into it and remain throughout the winter, eating nothing and moving not. Large
numbers sometimes find their way to the same place

**Hibernation
of Snakes.** and are often found coiled and twisted
together, thus giving rise to the many
stories of so-called "snake dens." If
the winter be an open one this hibernation, as it is
called, is often interrupted and the animal comes forth
from its retreat on some warm sunny day, thinking,
no doubt, that spring has come again. During an
excessive thaw the high water often finds its way into
the snake's resting place and many are doubtless
drowned while still torpid. Others escape and make
their way to a higher and drier spot. Thus on January 11, 1890, the writer found two species of garter

snakes beneath some fine driftwood, near the margin
of the overflowed bottoms, north of Terre Haute.
They had been driven forth from their winter retreat
by the high waters and had taken temporary refuge
beneath the drift.

On the first warm days of spring the sluggish blood
in the veins of the hibernating snakes begins to flow
more rapidly. Their bodily temperature gradually
rises. Demand for food and an irresistible desire to
mingle with others of their kind soon cause them to
move out and stretch their bodies in the warm sunshine and in a few weeks their summer haunts know
them as of yore.

The food of snakes is often the subject of much
conjecture among those persons who know the reptiles

**Food of
Snakes.**
only from an occasional chance meeting
with them. For example, some people accept literally the biblical state-
ment that they live upon dust, as the following inquiry
received by the writer will attest: "Is it necessary
that snakes have plenty of earth (or dust) to eat to
keep them alive? A friend of mine thinks that
snakes live largely on dust (or earth) but I do not
think so. Which is right? The dispute came up in
our young men's bible class on the reading of the following verse: 'And the Lord God said unto the serpent, because thou hast done this, thou art cursed above
all cattle, and above every beast of the field. Upon
thy belly shalt thou go, and dust shalt thou eat all the
days of thy life:' Genesis iii.—14."

Like other cold blooded animals snakes can fast for
a long time. In fact, in captivity, they have been

known to eat nothing for over a year although food
was frequently offered them. They need water, however, especially as their moulting time draws near.
What they eat depends to a great extent upon the
species, some preferring one kind of food, some
another. Like most men they are not fond of "cold
victuals" but prefer to capture their prey alive. Frogs,
tadpoles, small fish, young birds, field mice, rats and
especially insects and their larvæ, are their favorite
foods. When they take a notion to eat they believe
in "gittin' a plenty while they're gittin,'" provided
they have a chance. Thus, no less than seven large
leopard frogs, besides a mass of other material, were
once found in the stomach of a common water snake,
Tropidonotis sipedon (L.), which was dissected by the
writer on account of its aldermanic appearance.

Many snakes have also cannibalistic tendencies, so
that the accompanying illustrated "snake lie" has that
much for a basis.

Two instances of a snake's cannibalism have come
to the personal notice of the writer. Once, while
engaged in tracing a geological outcrop in the wilds
of Arkansas, he saw the top of a small bush shaking
gently to and fro. Investigating the cause of the
movement he found a half grown black snake with a
specimen of the summer green snake, *Cyclophis
æstivus* (L.), partially swallowed. The green snake
had been caught by the head and while endeavoring
to escape had wrapped its tail in a double coil about
the bush. The black snake had to suspend operations
when he had swallowed up to the bush, and was evidently awaiting the digestion of the part within his

body when discovered. This, no doubt, sounds like one of the "snake lies" referred to above, but the green snake, with the mark of the black one's teeth upon its body, was preserved in alcohol, and is yet in the writer's collection.

Fig. 6—An Illustrated "Snake Lie."

In the second instance a pupil brought in a king snake, *Ophibolus getulus sayi* (Holbrook), 12 inches in length, which had protruding from its mouth four inches of the tail of a common garter snake, *Eutainia sirtalis* (L.). The latter was 13 inches in length and the front nine inches of its body was within that of its captor. When taken the king snake tried to disgorge its prey but his gartership was too deeply lodged. The two were quickly consigned to a bottle of alcohol there to serve as a forcible illustration of the fact that the king snake is in reality both a king and a cannibal among its kind.

On account of their liking for field mice, insects and other vermin, each black snake on a farm is worth at least five dollars, and each garter snake one dollar, every year of its existence. In other words the damage which the mice, insects, etc., eaten by the snakes, would do, would amount to more than the sums mentioned. Both black and garter snakes are perfectly harmless, and yet as soon as one is seen eight farm boys out of ten, and almost as large a proportion of the farmers themselves, will procure a long club or a stone and mash the poor, defenseless snake's head to a jelly. Then if the snake be a large one, the exploit is bragged of all over the neighborhood. In the writer's opinion such an act is a cold blooded murder and the deed of a coward. As well might a giant brag of killing a dwarf as a man of killing a harmless and defenseless snake.

But few of the harmless snakes use their teeth as their chief means of defense. It is only when irritated or suddenly attacked that they will strike at a person, and the pain caused by their bite is no more severe than that produced by the bite of a mouse or the prick of a pin; the most serious result of the bite being usually the fright which timid persons sustain. Once in a great while, however, the bite of a harmless snake may cause a swelling of the organ bitten, and, one time in a thousand, may even cause death. This is due, however, not to any venom injected into the wound, but to a kind of blood poisoning brought about by some substance adhering to the teeth of the reptile, the victim being in a weakened physical condition and therefore very susceptible to such a result.

The bite of a mosquito has been known to prove fatal under similar conditions.

Their other means of defense and protection against their enemies are, however, numerous and interesting

Methods of Defense Used by Snakes.

to note. Almost all of them dart out the tongue when approached, seeking thus to terrify into retreat the aggressor. In this they are often successful, especially with mankind. Most other animals seem to know that the tongue is harmless and pay no attention to it. Some snakes, when molested, enlarge the body and so render themselves as formidable in appearance as possible. They do this, either by inflating with air to their fullest capacity, the long slender lungs, as does the spreading viper, whence the name "puffing adder" sometimes applied to it; or by flattening the body, by spreading out the ribs and then raising the scales, as does the common garter snake and the spreading viper. Certain species, when disturbed, force the air from their lungs with a hissing sound. This noise, no doubt, serves to frighten some of their enemies, but the expelled air is, in itself, wholly harmless.

A number of the larger snakes, among them the black snake, pilot snake, house snake and spreading viper, when alarmed, often try to imitate the peculiar rattle of the rattle-snake by vibrating the tail with great rapidity. If the vibrating tail happens to strike against some dead leaves the sound is very similar to that produced by the rattle-snake, and the writer, on hearing it, has frequently leaped back from a harmless snake thinking that he had been deceived as to the reptile before him.

No snake can, like the tree frog and the chameleon, change its colors to suit its surroundings; but many, during the summer time, frequent such places as accord most closely with their own hues. In this manner they not only lessen the chances of discovery by their enemies, but also increase their opportunities of obtaining food, as other animals, not perceiving them, will approach within striking distance. Thus, green snakes climb bushes and recline for hours on the slender branches among the leaves, waiting for insects and small birds to approach; and several of the smaller brown snakes remain for the most time among the dead leaves and grass, about logs and the roots of trees.

A few species, as the common garter snake and the spotted water snake, excrete a disgusting odor when handled and this no doubt serves to protect them from many enemies. Still others, when disturbed in their dreams on a bright spring morning, feign death or "play possum" as it is popularly put, remaining rigid and motionless as long as one stays in their vicinity, but seeking safety in flight as soon as they think themselves unnoticed. The above are a few of the many ways in which these reptiles seek to exercise their natural right of defending themselves. No one of the methods noticed is in the least degree harmful to man, and even if so, it must be remembered that whatever a non-venomous snake does is in self defense, as it is never the attacking party.

II.—THE SNAKES OF INDIANA IN PARTICULAR.

About sixty species of snakes inhabit that portion of the United States east of the Mississippi River. Of these the bite of but six is poisonous. Twenty-nine species and ten varieties are known to occur within the State of Indiana. Of these, four are poisonous, all the rest being perfectly harmless as far as the bite is concerned.

POISONOUS SNAKES.

With one exception the poisonous snakes of the State may be known from the harmless ones by the following characters. The head is broader than the body, flat and triangular, and has a deep pit on each side between the eye and the nostril, whence the name "Pit Vipers" which is sometimes given to the group. There are no solid teeth in the upper jaw, but on each side in front is a hollow poison fang which can be depressed against the roof of the mouth or erected at will. The canal in this fang connects with a duct or tube which leads to a poison gland on the upper side of the head. The poisonous liquid is separated from the blood by this gland and, when the serpent strikes, from four to six drops of it are injected through the duct and fang into the wound.

The liquid itself is tasteless, green to orange in color, and about ten times as heavy as water. Freezing, boiling, drying or treatment with alcohol does not affect its virulence. In man, as in most other animals, the poison causes great nervous prostration, lessens the number of heart beats per minute, and produces

something akin to blood poisoning. The best anti-
dote, as well known, is alcohol taken inwardly in the
form of whisky or brandy. This acts as a stimulant,
bracing up the system and enabling it to withstand
the depressing effects of the poison. When properly
attended to not more than twenty per cent. of the
bites of our poisonous snakes result fatally.

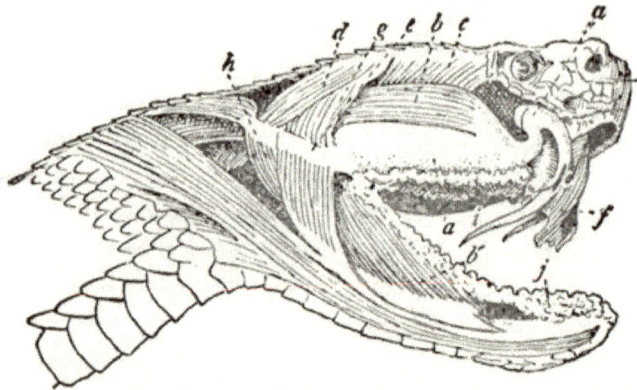

Fig. 7—Head of Rattle-snake, showing Venom Gland and Muscles.
a, venom gland; a', venom duct; f, sheath of fang; b, c, d, g and h, muscles;
i and j, salivary glands.

Hogs are seldom poisoned when bitten by a rattle-
snake or copper-head, as their fatty tissue absorbs the
poison and prevents it from entering the circulation.
Other animals usually die from the effects of the bite,
even the snake itself succumbing to its own venom
when it accidentally wounds itself. In other words,
the poison is a liquid secreted from the blood, which
becomes fatal on being introduced back into the very
same source.

Of the three "pit-vipers" occurring in Indiana, the
copper-head, *Agkistrodon contortrix* (L.), is readily dis-

tinguished from the other two by having its tail
devoid of a rattle and ending in a horny point. Sev-

eral species of harmless snakes are,
however, in many localities known as

Fig. 8—Tail end of
Copper-head.

"copper-heads" and are, therefore,
shunned as venomous. The most
common of these is the spreading viper or hog-nose
snake, which has a flat, triangular head, but which
lacks the "pit" between the eye and the nostril and
also the hollow poison fangs. Another

**The
Copper-head.**

marked difference is seen on the under
side of the tail where the plates or
scutes are, in the copper-head, mostly undivided,
whereas in the spreading viper they are divided on
the middle line. In color the copper-head is a chest-

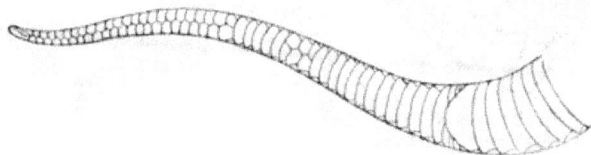

Fig. 9—Under side of tail of Southern Water-moccasin, a poisonous snake.
(After Stejneger.)

nut or hazel-brown, with numerous darker V-shaped
blotches along the back. Its head is a coppery-red,
whence the common name. It seldom, if ever, exceeds three feet in length, and its poison is less virulent
than that of either of the rattle-snakes. On the other
hand it is more justly feared than they, since it gives
no warning of an attack but strikes viciously and
repeatedly at whatever disturbs its repose.

°The Copper-head has similar undivided plates.

IV

COPPER-HEAD.
Ankistrodon contortrix (L.). (After Stejneger.)

The copper-head frequents for the most part rocky hillsides, especially those covered with timber and in the vicinity of water. Its young are born alive, and are few, seven to nine, in number. In the

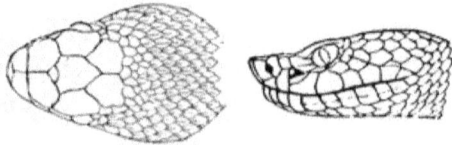

Fig. 10—Head of Copper-head, shown from top and side. (After Baird.)

early settlement of Indiana it was common in the southern half of the State, but at present one hears only of an occasional specimen; the most of those which are reputed as copper-heads, being found, upon examination, to be examples of some harmless species.

The banded or timber rattle-snake, *Crotalus horridus* L., reaches a length of six feet,* and a diameter of several inches. From the prairie rattle-snake it may

The Banded or Timber Rattle-snake.

be readily known by its having the top of the head covered with numerous scales instead of bony plates. In color it is yellowish brown with three rows of dark blotches, about twenty-one in each row, along the back between the head and the tail, the latter, in full grown specimens, being entirely black.

The rattle of this and allied species is composed of a series of flattened, horny rings joined rather loosely together, the terminal one, called "the button," being narrower than the others. The common belief that the age of the snake can be told by the number of

---

*A specimen in the State Museum from Arkansas measures six feet four inches, and its rattle is composed of thirteen rings and a button. Another from Clay County, Indiana, is five feet four inches in length and possesses eighteen rings and a button.

joints in the rattle is entirely erroneous, as many as four of the rings having been known to develop in a single year. Concerning this point Dr. L. Stejneger, the leading American authority on poisonous snakes, after mentioning the difficulty in overcoming the fallacy that "each ring on a rattle-snake represents a year of its life," says: "It ought not to be difficult to make people understand that the rattle is a delicate instrument which easily

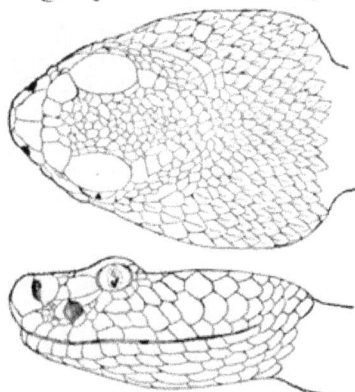

Fig. 11—Head of Banded Rattle-snake, shown from top and side. (After Baird.)

breaks; that old and huge rattlers are often found with but one or a few rings; that a variable number of joints are added each year, and that the production of a ring can be accomplished in the course of every two or three months."

In what manner has so unique an organ as the rattle developed? For what purpose is it used by the

Fig. 12—Separate joints of rattle of Banded Rattle-snake.
a, button; *h,* basal joint.

snake? These are questions which have been much discussed but are, as yet, unsolved. Some have likened the sound produced by the rattle to that made by the

BANDED RATTLE SNAKE.
Crotalus horridus L. (After Stejneger.)

harvest-fly, *Cicada tibicen* L., or by certain species of
grasshoppers, and have thought that the noise was
made to "decoy insect eating birds into the range of
the serpent's spring." Others have claimed that it
was a love call used in bringing the sexes together.
Still others have looked upon it as a "providential
arrangement to prevent injury to innocent animals

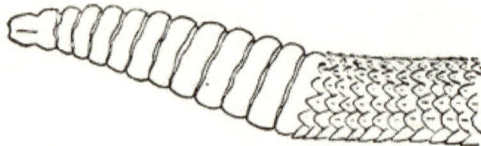

Fig. 13—Rattle of Banded Rattle-snake, (After Garman.)

and man." The most commonly accepted theory at
present is that it is used by the snake as a "means of
self-protection, serving the same purpose as the growl
of a tiger when threatened with danger. The snake
seldom sounds its rattle until it considers itself discovered, and not then unless it apprehends danger. It
throws itself in position to strike and says in unmis-
takable language: 'Look out for yourself, I am ready
for you. Your life, if you injure me.' If pushed
upon it makes its leap at its antagonist, and again
throws itself in position to renew the conflict, once
again sounding the note of defiance."* In making
its warning note the snake doubtless frightens away
many enemies which by experience have learned to
shun its presence. In this way it saves its venom, for
the use for which it is most evidently secreted—that
of quickly destroying or rendering helpless those
forms of life which the reptile needs for food.

* Henderson, J. G., American Naturalist, VI., 1872, 261.

The young of the timber rattle-snake are born alive.
They are seldom, if ever, more than nine in number,
and average at birth about eight inches in length. Its
food, in a state of nature, consists of rabbits, squirrels,
mice, and frogs, with an occasional bird, or harmless
snake to vary the menu. It usually lies in wait for
its prey. and when the latter comes in reach it strikes
at it with such rapidity that the motion can scarcely
be followed. Unless disturbed it ignores the presence
of man or of such animals as it does not wish for food,
and never follows such intruders with the intention of
attacking them.

Like the copper-head, the timber rattle-snake was
once rather common in southern Indiana, and doubtless occurred in small numbers in the northern half
of the State. At present it is known to occur only in
the broken, wooded portions of such counties as
Brown, Monroe, and Greene, where there are many
ledges of stone, on which, in summer, it can bask for
hours in the sunlight, and in whose crevices it can
find in winter a suitable abiding place. But here,
even, its numbers have become so few that the killing
of one is thought to be of sufficient importance for a
notice in the local newspaper, usually with a foot or
two of length added to the specimen.

The prairie rattle-snake or massasauga, *Sistrurus
catenatus* (Rafinesque), is, in general, smaller than the
The Prairie Rattle-snake. timber rattle-snake, seldom exceeding
three feet in length. The top of the
head is partially covered with horn-
like shields or plates, similar to those of the harmless
snakes. Its color is brown or blackish with seven rows

of darker blotches, 34 in each row, between the head and tail. Those specimens living in swamps and marshy places are often a uniform black in color.

In the wet prairies and marshes of northern Indiana the massasauga is yet found in small numbers, but is nowhere so abundant as it was a score, of years ago. No record of specimens taken in the State south of the National Road has come to the writer's notice.

Aside from its poisonous qualities, its habits are beneficial, as it feeds upon field mice and insects, and thus aids in keeping in check those pests. Its bite is considered much less

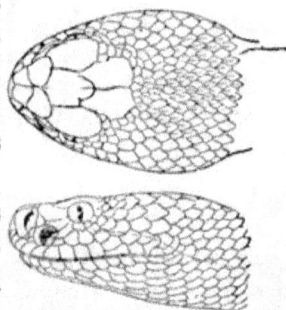

Fig. 14—Head of Prairie Rattlesnake, shown from top and side. (After Stejneger.)

venomous than that of the timber rattle-snake, one observer having asserted that it is scarcely more to be dreaded than the sting of a hornet. This might be true if the person bitten were so situated that immediate medical assistance could be obtained, but for persons living on a farm at some distance from a physician, the bite is always to be regarded as serious. The degree of danger from the bite of any of the poisonous snakes depends chiefly upon the size of the reptile, the amount of venom injected and the location of the wounded part, The larger the snake, the more venom it exudes and the deeper the fangs are driven into the body of the victim. If the wound is in such a part of the body that the poison is injected directly into the circulation the chances for recovery are small, no matter if the promptest of medical attention is obtained. Happily, such a bite does not often occur.

The fourth and last species of poisonous reptiles occurring in Indiana is the coral or bead snake, *Elaps fulvius* (L.). Unlike the "pit vipers," its head is but slightly distinct from the body. It lacks the pit be-

The Coral or Bead Snake. tween the eye and nostril, and the poison fang of the upper jaw is per-manently erect instead of movable at will. It is one of the most handsome of American snakes, being possessed of a slender body which is

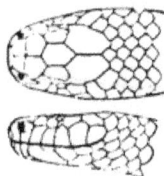

encircled by alternate bands of jet black and bright red, the latter color merging into yellow near the edges of the bands. The front por-tion of the head is black, while the hind portion is encircled by a band of bright yellow. The total length is less than 2½ feet.

Fig. 15—Head of Coral Snake, shown from top and side. (After Baird.)

In the southern States the bead snake is rather common and the extreme northern limit of its range is probably the southern half of In-diana and Ohio. In this State, but a single specimen, taken near Milan, Ripley County, has been recorded. Of its food habits but little is known, but that little goes to show that it is a cannibal, eating harmless snakes with evident gusto, since Dr. Stejneger records one as having swallowed a black snake as long as itself, before it had fully digested a garter snake taken at a previous meal. Much discussion has taken place concerning the ability of the coral snake to inflict dan-

gerous bites. **There is** little doubt, however, but that
it injects a true poison which sometimes causes fatal
results, but on account of the smallness of the serpent's
mouth and the shortness of its poison fangs, the wound
must be inflicted on the more exposed portions of the
body, as the fingers or toes. It is hoped that persons
in the southern half of Indiana will, in the future, be
on the lookout for this snake, that a more definite
account of its range in the State may be put on
record.

HARMLESS SNAKES.

For convenience the twenty-five species of harmless snakes known to occur in Indiana may be classed
according to color, habits, etc., into seven groups.
Three of the reptiles are seldom found far from ponds
or streams, and hence may form

Group I.—The Water Snakes.

Two of these are quite similar in habits and appearance, the ground color varying from ashy to
brown, with a row of thirty or more darker spots on
the back between the head and tail, and a row of
smaller, similar spots on each side. Scientists distinguish them, however, by the difference in the number
of rows of scales on the back, giving to the rarer one,
which has 27 rows, the name of "diamond water
snake."

The other one, which has but 23, rarely 25, rows, is
Tropidonotus sipedon (L.), one of our most common
snakes and popularly known as the "water snake,"

"spotted water snake" and "water moccasin." About the larger ponds and streams, especially those of southern Indiana, it grows to a large size, reaching a diameter of three inches and a length of five feet; but in the central and northern parts of the State specimens more than four feet long are seldom seen. It is usually given a wide berth, as seven people out of ten believe that its bite will cause certain death. This belief is no doubt caused by the fact that the

The Spotted Water Snake. "water moccasin" or "cotton mouth" of the southern States is a poisonous snake, and the common names of the two have become confounded. Our water moccasin

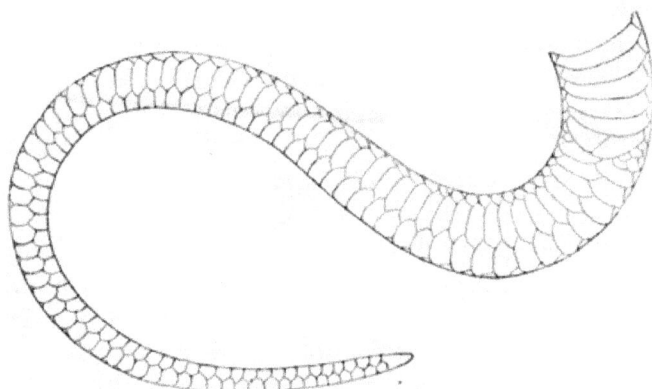

Fig. 16—Under side of tail of Spotted Water Snake, showing divided plates or scutes. (After Stejneger.)

has no fangs whatever, and its bite is never more serious than that of a mouse. It is partial to still waters of considerable depth, and seldom frequents streams that have not a bottom of deep, soft mud, in which to take refuge when pursued, and in which it buries itself deeply during the winter. It delights in the piles of driftwood which collect about such pools, and on a midsummer day three or four may be seen stretched out on the same log, evidently enjoying the sunshine and awaiting the near approach of their favorite prey, the leopard and bull frogs. Sometimes another snake, swimming too near, pays with its life the penalty of its rashness. Minnows also, doubtless, form a large proportion of their food, and an instance is on record where an individual of this species was surprised with a pickerel a foot long in its mouth.

The young of the water snake and its near allies are hatched from eggs either within the body of the mother, or very soon after the eggs are laid, and as many as 33 have been recorded as belonging to a single brood. Several color varieties of this water snake occur in Indiana, one of which, a uniform blue-black above and reddish beneath, is known as the "black water moccasin," *Tropidonotus sipedon erythrogaster* (Shaw).

The diamond water snake, *Tropidonotus rhombifera* (Hallow.), is, as above mentioned, a distinct species, known by its 27 rows of strongly keeled scales; i. e.,

The Diamond Water Snake. scales with a ridge extending lengthwise of the center of each, and by the squarish brown spots on the back alternating with those on the sides and connecting with them at the angles. Several specimens of this snake from southern Indiana are in the State Museum, one of which, from Morgan County, was labeled "Copperhead—*Trigonocephalus contortrix*—A poisonous American Serpent, called also copperbell and red viper."

The diamond water snake reaches a larger size than the common water snake, and the two are often confounded by observers. Its habits are essentially the same as in that species. Both strike viciously when disturbed, and exhale a very disagreeable odor when handled, this being, probably, their most efficient means of defense.

The third species of water snake found in the State is the queen or leather snake, *Regina leberis* (L.), a much smaller and more slender reptile than either of the last two, seldom exceeding two feet in length.

The Leather Snake. Its scales are keeled and occupy 19 rows, while its color is olive brown with three narrow black stripes on the back and a yellowish band along the side. Beneath, it is yellowish with two brown bands which lie close together and reach from the head to the tail.

Along the rapid flowing streams of central Indiana this is a very common snake, and it probably occurs throughout the State. It frequents, for the most part, shallow running water, gliding gracefully among the stems of the water willow, *Dianthera americana* L., and other aquatic plants; and, when pursued, taking refuge beneath some flat stone, or the piles of driftwood along the shore. It is never, as far as my observation goes, found at any distance from water, and its food consists mainly of small "peeper" frogs, young toads and minnows.

Group II.—Black Snakes.

To this group belong four species of our largest snakes. The ground color of each of them is black, and the common name of "black snake" is indiscriminately applied to them. However, a little practice soon enables one to distinguish them apart. The most common of the four and the only one to which the name rightfully belongs, is *Bascanion constrictor* (L.), a snake which is a uniform deep black above, paler beneath, and has the scales on the back perfectly smooth and in 17 rows. The young, up to the third moult, are very different in color from the adult, being

The Black Snake or Blue Racer.

olive brown with numerous large, darker colored spots along the sides. When they are about two-thirds grown, the hue is of a bluish shade, and they are then commonly known as "blue racers"; most people believing them to be an entirely different snake.

More "lies" have been told about this snake than any other one in existence. It "charms birds," "sucks cows," "steals eggs," "drinks the milk in the milk houses," "kills a rattle-snake by pulling it in two," and does fifty other deeds that no snake on earth ever did or ever will do. One thing, however, it can do, and do well, and that is to turn tail and run when approached, seeking a shelter with "that celerity of movement no other creeping creature can obtain."

The black snake feeds principally upon rats, mice, crickets, grasshoppers and beetles, and may occasionally swallow another snake or a small bird for dessert. However, the good

Fig. 17—Head of Black Snake. (After Baird.)

that they do far outweighs the bad, and yet every

year their numbers are becoming less, for "death to the black snake whenever and wherever found" seems to be the watchword of all boys and most men. Another cause for their lessening numbers is undoubtedly the rapid disappearance of the old Virginia rail fences, beneath the bottom rail of which they were formerly sure of a safe retreat from all attacks.

At certain seasons of the year, as in spring when mating, and in late autumn, when seeking a hiding place for the winter, the black snake is vicious, hissing and striking at a person who is several yards away. At such a time it will occasionally pursue a person whom it recognizes as more cowardly than itself, and in this way has probably gained the name of the "blue racer." When seized by the neck it quickly throws a double coil about a person's arm and gives a grip with its powerful muscles which the captor has no little difficulty in breaking. The stories which one often hears of its attacking persons and squeezing them to death are wholly without foundation. The young are hatched from eggs which are usually deposited in soft earth or the humus of decayed wood. These eggs are an inch and a half long by an inch in diameter, and covered with a tough, thick skin. According to Dr. Hay, as many as nineteen eggs are laid at a time, and from one ready to hatch he took a young racer ten and one-half inches long.

As to the many stories concerning the size to which the black snake grows, mention has been made on a previous page. A little over six feet is doubtless their maximum length, yet they are often said to have been seen eight and even ten feet long. In regard to the length of snakes generally, Dr. C. C. Abbott in one of his charming books has well said: "that with timid people, a great deal depends upon the direction in which the snake was moving at the time it was seen. As an old, observing friend once said to me, ' When snakes come towards folks, every foot looks a yard long.'"

One of the largest snakes found in Indiana is the "pilot snake" or "black racer." It is often confounded with the true black snake **The Pilot Snake.** or "blue racer," but has the scales in 25 to 29 rows instead of 17, those along the middle of the back being obscurely keeled. In the place of being uniformly black above, it usually has some of the scales white-edged, thus causing some fine white mottlings on the upper side.

The pilot snake at times grows to be six and one-half feet long and the body is always much thicker than that of a black snake of the same length. It frequents dry, open woods and thickets, and more often than any other of our snakes is seen in bushes, and even in the tops of tall trees where it has climbed by following the depressions in the rough bark. Although its bite is harmless, yet it is, probably, our most injurious snake on account of its liking for small birds, which form one of its principal foods, and for which it undoubtedly lies in wait in the bushes and trees. Field mice and insects form also a large portion of the food, so that it makes up in part for its depredations among the birds.

The young of the pilot snake are hatched from eggs which are deposited by the mother in such places as hollow stumps and close alongside old logs. The young, for the first year or more of their lives, are ashy gray with about 45 square, chocolate blotches on the back, and a row of alternating smaller blotches along each side. There is also a dark band between the eyes, and the foremost spot on the back is forked, each division extending a short distance onto the head. One of these young, 16 inches in length, which contained a large shrew, partially digested, was taken June 11, 1894.

A pilot snake over five feet in length was once kept by the writer in a vacant room with a great horned owl, some turtles and salamanders. It was supposed that the size of the owl, which is one of our largest birds of prey, and has a very strong beak and talons, would prevent a conflict, and that a "happy family," equal in interest and peaceful inclinations to any seen in a menagerie, would result, but events proved otherwise. One night a strange noise was heard in the room, and on investigating its cause it was found that a "struggle for existence" had taken place between the two leading members of the family. Whether the snake attacked the owl or the owl the snake was never known, but the snake proved itself the "fitter in the struggle," and quickly squeezed the life out of the owl by wrapping two coils tightly about it. Perhaps the snake would, if let alone, have attempted to swallow the owl, but a desire to secure the latter in as good a condition as possible for a permanent specimen led to its immediate removal from a literal embrace of death.

A third snake, which reaches a length of four feet or more, and which in the country usually goes also by the name of "black snake," is the king snake, *Ophibolus getulus sayi* (Holbrook). It is less common

The King Snake. than either of the last two mentioned. From them it may be known by the scales being smooth and in 21 rows. Many of the scales have a small yellowish spot in the center, and in young specimens these spots often unite to form cross lines on the back. These lines sometimes fork on the sides and divide the black of the back into large blotches. It will thus be noted that the young of many black or dark colored snakes are always spotted, and that as they grow older and shed their skins a number of times they gradually grow darker, until finally they become almost wholly black. This has, in the past, been the cause of much confusion in the naming of the reptiles, many of the young having been thought to be distinct species.

The king snake frequents open woodlands and the borders of moist thickets, feeding upon mice, moles, toads, salamanders, and, as noted near the beginning of this paper, upon such other snakes as it can conveniently swallow. It is a very active reptile, but in general mild and inoffensive in its habits. When cornered, it will strike rapidly and viciously, causing the timid person who has suddenly come upon it to beat a hasty retreat.

In some of the southern States, where it is more common than in Indiana, it is reputed to wage a successful warfare upon the rattle-snake, and hence received its common name. A prominent writer and

generally accepted authority on snakes, evidently trying to excel some newspaper reporter in the production of a snake story, avers that: "By suddenly
springing upon and encircling the rattle-snake with
its coils, the king snake soon squeezes the venomous
reptile to death. Then, commencing at the head, the
victor swallows the rattler whole."

The last species which belongs to the group of black
snakes is the horn snake, *Farancia abacura* (Holbrook).
It is said to be rather common in the southwestern
States, but in Indiana has been taken only near
Wheatland, Knox County. It is a handsome species,
as handsome goes among snakes, being blue-black
above with about sixty squarish red
The Horn   spots on the sides. These, in some
Snake.   specimens, extend nearly to the middle
of the back. Beneath, it is red, blotched with black.
The scales are smooth and in nineteen rows.

Having never met with this species alive, I can say
but little of its habits. According to its first describer,
it is shy and lives in swampy ground and damp thickets. It reaches a length of four feet or more. People
in the southern half of the State should be on the
lookout for it, and if a specimen is secured it should
be sent to the State Museum or presented to some
school which will preserve it for future reference.

Group III.—Spotted Snakes.

Among the harmless snakes occurring in Indiana,
which are usually found at some distance from water,
are four species of medium or large size which are

distinctly spotted throughout their entire lives. One
of the most handsome of these, and one quite frequently met with in dry, upland woods, and about
country houses and barns, is the "spotted adder,"
Ophibolus doliatus triangulus (Boie.), "house snake,"
"milk snake," "thunder and lightning snake," or
almost anything else one may wish to
The House or call it, as it is a creature of many
Milk Snake. names, of which the above are the
most common. It varies much in color, but is
usually grayish with three rows of brick-red, black-
bordered blotches on the back and sides; the larger
ones saddle-shaped and alternating with the smaller,
the latter being often wholly black. There is usually
a light colored arrow-shaped spot back of the head,
while beneath, the body is checkered with black and
creamy white. The scales are smooth and in 21 rows.
The young are hatched from eggs which are about
two inches long and a little more than an inch in
diameter. During the first year of their lives they
are often found beneath the loose bark of logs and
stumps, where they are doubtless seeking the crickets,
cockroaches and other insects which have there their
abiding places.

The house snake sometimes reaches a length of
four feet, and when disturbed resents only by darting
out its forked tongue and giving an occasional vibra-
tion of its tail. Its usual food consists of mice, rats,
and such unfortunate toads as happen in its way,
except in grasshopper season, when it feasts to its
stomach's content upon those festive insects. It is
often found about spring houses where milk is kept,

presumably in search of frogs and salamanders which frequent such damp localities; but the owner of the milk usually asserts that it drinks that lacteous fluid and hence gives it the name of "milk snake."

In rare instances double-headed specimens of this snake have been taken. The writer has seen one in which the two heads were each about three inches long and then united into one body.

According to scientists, the house snake is only a color variety of the red snake, *Ophibolus doliatus* (L.). The latter is red or scarlet and has 20 or more pairs of black rings, each pair enclosing a yellow spot. Its habits are essentially the same as those of the house snake, but it is much less common and seldom grows above two feet in length. It has been taken in a number of localities in southern Indiana, but in the north only the more spotted variety has, as yet, been found.

Another spotted reptile closely allied to the house snake is the chain snake, *Ophibolus calligaster* (Say). Its smooth scales are in 25 rows, and it has about 60 squarish chestnut-colored blotches along the back which alternate with smaller rounded **The Chain Snake.** spots along each side, the ground color being olive gray.

The range of the chain snake is western, and but a single specimen is so far known from Indiana. It was taken by the writer from open woods just east of Terre Haute in Vigo County, and is about three feet in length. Nothing distinctive is known of its habits, though in Illinois it is said to frequent prairies, where it doubtless lives mainly upon small mammals and insects.

The fox snake, *Coluber vulpinus* (B. & G.), is a third distinctly spotted snake which is found occasionally in Indiana. The ground color is light brown or grayish-yellow, and there are about 60 chocolate-colored spots across the back, which

The Fox Snake.

alternate with smaller ones on each side. It may be known from the chain snake by its having nine or more of the 25 rows of scales keeled. The under surface is yellowish, with large squarish blotches of black.

The fox snake feeds upon the smaller mammals, as half-grown rabbits, mice and ground squirrels. Like most snakes it is irritable and vicious when surprised immediately after swallowing its prey, evidently fearing that the intruder will cause it to disgorge the latter and so deprive it of its dinner.

The fourth species of spotted snakes occurring in the State is the corn snake, *Coluber guttatus* L., a reptile of southern range, which has been taken at a few localities in southern Indiana. It is thought by some

The Corn Snake.

to be only a variety of the fox snake, but the scales are in 27 rows and the ground color is brick-red instead of gray. The dark blotches are also fewer, being seldom more than 45.

Not having seen the corn snake alive I can say nothing of its habits. Holbrook says that in North Carolina "it is found about the roadsides early in the morning or at the dusk of evening, unlike most snakes concealing itself during the day. It is very gentle and familiar . . . at times entering houses, and is, according to Catesby, a great robber of hen-roosts."

Group IV.—Striped or Garter Snakes.

Four slender bodied reptiles, whose general color consists of three light stripes on a darker ground, with sometimes intervening darker spots, and with the lower side unspotted, belong to this group of harmless snakes. The scales of all are keeled and in 19, rarely 21, rows.

Of these the ribbon snake, *Eutainia saurita* (L.), is much more slender and graceful, and withal a hand-

The Ribbon Snake.

somer species than any other, its color being a dark, glossy, chocolate-brown, with the three stripes of a bright greenish yellow. It reaches a length of three feet or more, and its favorite haunts are damp thickets and the borders of streams and ponds, where, on the first bright sunny days of spring, several may sometimes be found in close proximity. It often takes to water to escape its enemies, swimming with graceful curves of its long slender tail, but it is by no means aquatic in its habits. The food of the ribbon snake is chiefly insects and the small cricket frogs, its body being too slender to encompass many of those larger forms of life in which other snakes delight. A stouter bodied, darker colored variety of this species is sometimes called Fairey's garter snake.

The common garter snake, *Eutainia sirtalis* (L.), is the most common reptile in the United States, and at the same time one of the most variable. Four varieties are known to occur in Indiana as follows:

(*a*)   Grass snake, *E. s. graminea* Cope; color, green, lacking both stripes and spots, except a small black spot near the end of each ventral plate.

(*b*)   Spotted garter snake, *E. s. ordinata* (L.); lacks the stripes, but has three distinct rows of square dark spots, on each side between the head and tail; and also a small black spot at the end of each ventral plate.

(*c*)   Red-sided garter snake, *E. s. parietalis* (Say); has the stripes present, yellow or greenish, and a row of brick-red spots alternating with a row of darker colored spots along the sides.

(*d*)   Common garter snake, *E. s. sirtalis* (L.); has the stripes present but faint and narrow, with three rows of indistinct dark spots on each side.

These varieties are named in the order of their relative abundance in Indiana, the grass snake being scarce, while in the course of the summer more of the red-sided and common garter snakes are probably seen than of the individuals of all other species combined. They are to be found anywhere, but prefer the vicinity of water, where frogs and other snake food is most abundant. When teased, they flatten the body, elevate the scales, hiss and exude a disgusting odor. Sometimes they strike viciously and may even draw blood with their needle-like teeth, but their bite is less harmful than that of a mosquito, as it causes no swelling or after pain. The young of the garter snake are born alive in late summer or early autumn, and their number is legion. Dr. J. Schneck, of Mt. Carmel, Illinois, has recorded* the taking of

The Common Garter Snake.

*American Naturalist, XVI, 1882, 1008.

78 from three to seven inches in length from the body of a single female 35 inches long. Other observers have noted from 35 to 80.

This snake has many enemies, among which may be mentioned owls, hawks, hogs, skunks, ducks, turkeys, other snakes, and last but not least, the small boy with a big club. Feeding as it does mainly upon insects and the smaller injurious mammals, the good that it does far outweighs the bad. In the future, therefore, let its presence in the dooryard be looked upon with favor, and let the hand be stayed that in the past has ever been raised against it.

The Racine garter snake. *Eutainia radix* (B. & G.), has been taken in a few localities in western Indiana. Prof. E. D. Cope says that it is the prevailing garter snake of the western plains, ranging from the base of the Rocky Mountains on the west to the eastern limit of the prairies in Indiana on **The Racine Garter Snake.** the east. Its scales are usually in 21 rows and are prominently keeled so that the reptile is very rough in appearance. In color it is dark olive brown with the lateral stripe on the third and fourth rows of scales instead of on the second and third, as in the common garter snake. The stripe on the back is bordered with black and there are two rows of dark spots on each side between the stripes. Below the lateral stripe there is also an additional row of small spots. The average length is about two feet.

The habits of the Racine garter snake are essentially the same as those of the more common species. Living as it does mainly in prairie regions it is largely

preyed upon by hawks and in turn preys upon the
smaller mammals. These live upon the insects, which
gain their sustenance from the grass and grain. Thus,
by tracing back the food of each, we realize the force
of the old saying: "All flesh is grass," and know
that the plant must have existed before the insect,
reptile, bird or mammal made its appearance upon
earth.

Butler's garter snake, *Eutainia butleri* Cope, was described from a specimen taken near Richmond, Indiana. But two additional specimens have since been
secured, one at Waterloo, Dekalb County, and the
other near Turkey Lake, Kosciusko
County. It is distinguished by the
yellow, black-bordered lateral stripe
covering *three* instead of two rows of scales, and by
its head being much smaller and more conical than in
the typical garter snakes. The eye is also propor-
tionally smaller than in any other of the more
common species. Nothing distinctive of its habits
is known, and additional specimens are greatly to be
desired.

**Butler's
Garter Snake.**

Group V.—Green Snakes.

Of all our reptiles not one can exceed in beauty
and gracefulness the "green" or "summer snake,"
Cyclophis æstivus (L.) It possesses 17 rows of keeled
scales and in color is a uniform bright green above
and light yellow beneath. With a body remarkably
slender for its length, which at times reaches thirty or
more inches, it is an object which delights the eye of

5

the nature-loving rambler whenever he is so fortunate as to meet with it. The favorite haunts of the summer snake are rocky hillsides, especially those in the vicinity of running water. Oftentimes, too, the seeker after wild berries is needlessly frightened by seeing one reposing on the bushes within a few inches of his out-stretched hands. One of these snakes, kept in captivity by the writer, often rested on the posterior half of its body, and, raising the front half almost vertically, it would remain rigid and motionless for half an hour at a time. In its wild state such a habit, if practiced, would render it, for the time being, very secure against such enemies as were guided only by the sense of sight, and would allow the near approach of such small animals as the snake subsisted upon.

The Summer Snake.

The habits of this species were excellently portrayed by Dr. Holbrook, who wrote of it as follows: "The summer snake is perfectly harmless and gentle, easily domesticated, and takes readily its food from the hand. I have seen it carried in the pocket or twisted around the arm or neck as a plaything, without ever evincing any disposition to mischief. In its wild state it lives among the branches of trees and shrubs, shooting with great velocity from bough to bough, in pursuit of the insects which serve as its nourishment. Its green color, similar to the leaves among which it lives, affords it protection against those birds which prey upon it."

Another green snake, as handsome as the one last mentioned, and distinguished from it only by its scales

being smooth and in 15 rows, is the smooth green
snake, *Liopeltis vernalis* (DeKay). It is much less

**The Smooth
Green Snake.**
common in Indiana than the summer
snake, and is found usually in the tall
rank grasses which grow about the
margins of marshes and swamps. There it lies in wait
for the green grasshoppers and katydids which are so
abundant in such a place. Except in these grasses it
is seldom seen higher than the surface of the ground.
Its eggs, an inch and a half long, were found by one
observer beneath the bark of an old stump, and one
young snake, just hatched, was five inches in length.

Group VI.—Small Brown Snakes.

The snakes heretofore mentioned comprise the
giants of the family as found in Indiana. We shall
now deal for a time with the dwarfs. Six of the 29
species occurring in the State, when full grown, seldom, if ever, exceed 16 inches in length.

The most common of the six is known as DeKay's
brown snake, *Storeria dekayi* (Holbrook). It is grayish-
brown in color, paler beneath, has a narrow pale band

**DeKay's
Brown Snake.**
along the back and a dark spot behind
each eye. The scales are keeled and
in 17 rows, and the average length is
about one foot. This is one of the first snakes seen
in the spring, a specimen having been taken by the
writer as early as April 2nd. Like many other species
it is then usually found close to water, and doubtless
breaks its long fast upon one of the small "cricket"
or "peeper" frogs, whose shrill and countless voices

make the welkin ring on just such days as tempt the
snake forth from its winter's retreat. As these frogs,
small as they are, are fully twice the diameter of the
snake, it is doubtless with much effort that this first
spring meal reaches its final resting place in the latter's stomach. Later on in the season young and tender grasshoppers and crickets furnish them a bountiful
repast, and it is even affirmed of them by Abbott that
"they are excellent fishers, and gliding through the
water with marvelous celerity, they catch minnows
and young pike in large numbers." The young of
this species, as well as those of the next, are hatched
from eggs within the body of the mother, and number from eight to fifteen.

Another snake which, from above, closely resembles the last mentioned in color and size is Storer's
brown snake, *Storeria occipitomaculata* (Storer). Its

Storer's Brown Snake. scales, however, are in 15 rows, and on
turning it over, a difference can be
readily seen as it is a deep salmon-red
beneath, whence it is often called the "red-bellied
brown snake." Its usual home is beneath logs and
stones where it feeds upon crickets, myriapods, slugs,
earth worms and other crawling creatures.

On one occasion while driving in Vigo County the
writer saw a chicken running along the roadside with
a wriggling snake in its bill. After a sharp chase of
the fowl through a rail fence and a blackberry patch,
its prey was dropped and proved to be a fine specimen of Storer's snake. As soon as it found itself
free it wrapped its tail about a small bush and when
approached flattened itself very much after the man-

ner of a spreading viper. The row of brown dots
bordering the pale band along the back
then became much more prominent than
they were when the body resumed its
normal shape.

Most snakes have a head larger than,
and distinct from, the body, but there
are two occurring in Indiana which have
the head indistinct, it being at the base
of the same width as the body whence
it tapers gradually to a dull point. The
more common of the two is the ground
snake, *Carphophiops amœnus* (Say).

The Ground or Worm Snake. Twelve inches is its maximum length, and on ac-
count of the small size
and the tapering head it is often called
the "worm snake." In color it is a
glossy chestnut-brown above
and red or pinkish below,
while the scales are smooth
and in 13 rows. It lives, for
the most part, coiled up beside
or beneath rotten logs, among
dead leaves, and about the
roots of trees. In such places
it readily makes its way, forc-
ing its sharp muzzle into narrow crevices with much muscular strength. Such surroundings also harmonize with its
colors, and crickets and other

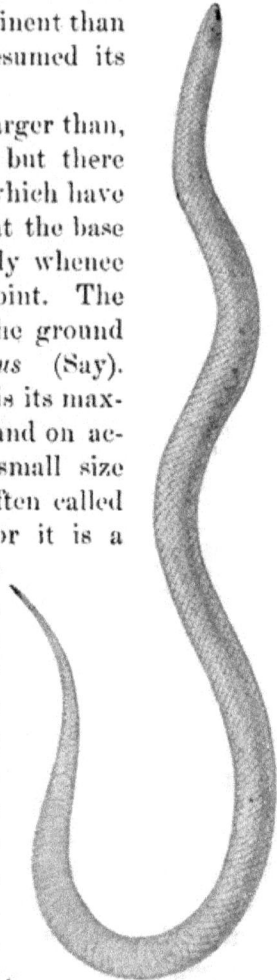

Fig. 18—Ground or Worm
Snake.

snake delicacies are there plentiful—two conditions of life, which, if a snake possess, fully satisfy it here below. It is perfectly harmless, not being able to open its small mouth sufficiently wide to bite a person if it would; although by twisting about one's wrist or finger it may cause an involuntary snaky shudder to creep up his back.

The second species with the conic head indistinct from the body is Virginia's snake, *Virginia elegans* Kennicott. It is light olive brown above and yellowish beneath, while scattered over the **Virginia's Snake.** upper surface are numerous small black dots, resembling points made by a fine pen. The scales are in 17 rows, very narrow and faintly keeled. In Indiana Virginia's snake has been recorded only from Brown County. Its habits are unknown, but presumably the same as those of the worm snake, since it frequents similar localities.

Two other small snakes remain to be mentioned, and although neither is strictly brown yet they will be treated of in this connection. One is the "ring-necked snake," *Diadophis punctatus* (L.), a handsome little reptile, blue-black above, pale **The Ring-Necked Snake.** orange below, and with a conspicuous yellowish ring about the neck. The smooth scales are in 15 rows, and on the outer end of each of the ventral plates there is usually a small black spot, while a median row of similar spots is sometimes present on the under side between the head and tail.

Growing to a length of a foot or more, the ring-necked snake is usually found beneath the loose bark

VI.

I. RING-NECKED SNAKE.

II. DeKAY'S BROWN SNAKE.

III. KIRTLAND'S SNAKE.

of a fallen tree or under a chunk on some dry hillside. In such a place it feeds upon those insects which come readily to hand and, when disturbed, seeks to defend itself only by exuding a disagreeable odor.

More common than the above, especially in central Indiana, is Kirtland's snake, *Tropidoclonium kirtlandi* (Kennicott). It also is very prettily marked, being light reddish brown with two rows of large round **Kirtland's** dark spots on each side, while beneath **Snake.** it is a bright salmon red, with a row of black spots along the margin of the ventral plates. The scales are keeled and in 19 rows.

This snake evidently feeds at night, for of 20 or more taken by the writer, all were found coiled up beneath logs or stones, seemingly half asleep, as they were very sluggish in their actions even after their chosen shelter had been rolled from above them. Its only show of self defense is a habit of flattening itself so that it becomes very broad and thin. It then strikes viciously for several times, when, seemingly satisfied with its show of resistance, it coils itself up and quietly eyes the intruder. Toads, frogs and insects comprise its food, and the young are born alive.

Group VII.—The Vipers.

And now we come to the last, the ugliest and the clumsiest of them all, the "hog-nosed snake" or "spreading viper," *Heterodon platyrhinus* Latreille. Much has been said of it on the previous pages but it deserves more than a passing mention, as none other of our snakes can hiss more loudly, spread more flatly, or snap more fiercely; and none other, as commonly met with, is as much feared as this. And yet, in spite of all these threatening actions,

The Spreading Viper.

its bite is perfectly harmless, as the writer knows by experience. It has no sign of a poison fang and no duct connecting with a poison gland. Aside from its actions the spreading viper may be known from other Indiana snakes by having the snout brought to an edge along the sides and to a point in front, and then turned up so as to resemble a pointed dirt shovel. In color this snake is either uniform black, or a yellowish brown with about 28 darker blotches on the back and sides. The spotted form is the one most usually

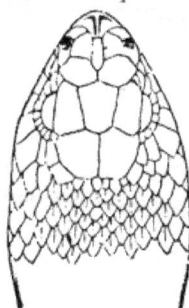

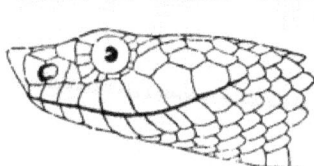

Fig. 19—Head of Spreading Viper, shown from top and side.

seen but the other in this vicinity is not rare. The scales are keeled and disposed in 23 or 25 rows. It sometimes reaches a length of three and a half feet and is then possessed of a thick and heavy body.

A closely allied form, both in structure and habits, is the "sand viper," *Heterodon simus* (L.), which inhabits the southern States and has been taken in Indiana at New Harmony and Brookville. It is distinguished from the spreading viper by having the central plate of the head surrounded by five to ten small plates. It seldom exceeds two feet in length and the scales are sometimes in 27 rows.

The young of the spreading viper are hatched from eggs which are buried in loose soil, sand or the humus of decayed logs. The eggs are about 1¼ x ¾ inches in size and covered with a tough, yellowish membrane. As many as 27 are known to belong to the same batch. When just hatched the young are about 8 inches in length and are ready to hiss, flatten the body and strike viciously whenever teased.

A singular habit possessed by the spreading viper is that of sometimes feigning death when disturbed. This is more often indulged in in early spring soon after they have left their winter retreats and while they are seeking their first spring meal or choosing their future mates. On March 23, 1893, a black specimen was found coiled up in some dead leaves in an open place in the woods. On being teased it went into a "fit," turning on its back with its mouth wide open and its tongue protruded at full length. Whenever it was turned right side up it would immediately turn on its back again. If held right side up with a stick it would squirm vigorously and endeavor to turn over. It was left lying bottom up, but on my returning to the spot a half hour afterward had disappeared.

On another occasion a black and a spotted one were found in company and when disturbed they opened wide their mouths, turned on their backs and coiled and twisted about in a very rapid and curious manner for about five minutes, when they became quiet and apparently lifeless. During all these contortions they had remained on their backs, and when they became quiet and were turned over they would immediately turn on their backs again, but otherwise gave no signs

of life, even at the end of an hour's time. According
to Dr. Hay, the newly hatched young, when teased,
will undergo the same contortions, and will lie perfectly still on the back until they think they are unobserved when they will turn over and slyly creep away.

The favorite resort of the "spreading viper" is a
sandy hillside with a southern exposure, or the borders of an open or cultivated field. In such a locality
their principal food consists chiefly of noxious insects,
and hence they, as well as all other harmless snakes,
should merit the protection of man instead of being
forever an especial target of his insane desire to kill
all objects beneath him in the scale of life.

TO A GARTER SNAKE.

Thou art humble
And content to crawl
Upon the lap of earth:
To seek thy food without the brawl
And strife, which others,
Far above thee in the scale of life,
Do use.
Thou art harmless,
And yet upon thy head
Has ever been a curse unmerited;
Making of thee a shunned, polluted thing,
Although thou art possessed
Of neither fang nor sting.

And even now,
In this enlightened age,
Man sees thee but to spurn
And strike at thy poor form;
And on the printed page
Thy name is seldom uttered
Save with words of scorn.

BLUE-GRAY GNAT-CATCHER AND NESTS.

The nests are the ones mentioned in the text, the one on the right being the double one.

A FEATHERED MIDGET AND ITS NEST.

Next to the humming-bird, the blue-gray gnat-catcher, *Polioptila cærulea* (L.), is the smallest bird nesting in Indiana. It is a summer resident, arriving from the south about the tenth of April; the date of its arrival in Vigo County for five successive years having been April 10th, 11th, 11th, 10th and 10th, respectively, showing that it can judge the day of the year almost as well as some beings higher in the scale of animal life.

The total length of the bird is but 4½ inches, and of this, 2¼ inches is tail. The color is an ashy blue, brightest on the head; the male with the forehead and a line over the eye, black.

By the time the gnat-catcher arrives, insects of various kinds are plentiful and its season's work of lessening their ranks at once begins. On April 18, 1897, I watched for an hour four of these birds in their ceaseless insect-seeking movements. They were in a thorn tree and I in the angle of an old rail fence, less than a dozen feet away. Flitting from twig to twig; turning their heads now this way, now that; peering first on one side of a branch and then on the other, they kept up their eager quest. Every few moments one would dart out to one side of the tree and catch an insect on the wing. Once, while endeavoring to catch a rapidly flying beetle, one of the feathered sprites turned a complete somersault in the air. Again, another flew close to the ground within a foot of my reclining form, caught a small moth hovering above the grass, and then darting back to its perch gave the insect two or three whacks against a branch, either to kill it or to straighten it so that it could be easily swallowed, and then gulped it down.

One caught a moth as large as a cabbage butterfly, and struck it on a limb for several seconds. Four times the luckless insect got away but was each time recaptured in short order, and was finally, after repeated shakings and beatings, swallowed, wings and all—the bird stretching and gaping for some little time thereafter, much as does a hen which has swallowed a tight fit for her œsophagus. It seems, therefore, that if the insect be small it is swallowed as soon as caught, often before the bird reaches its perch. If large, the beating on the limb or other resting place of the bird takes place.

When insect life, in and close about the thorn tree became, for a time, scarce, one or two of the birds would fly to the near-by fence and flitting along its angles would sometimes be rewarded by starting up an unlucky insect which would be instantly nabbed. Again returning to the thorn they would fly to a papaw, on whose large, velvety, expanding buds small bees and flies were plentiful; but the thorn seemed their favorite base of operations, and to it they invariably returned.

The long tail of the gnat-catcher serves it admirably as a rudder, and in the stiff breeze which was blowing, was bent now this way, now that, to preserve

the balance of the owner. The tail of the male bird
is darker than that of the other sex. When in flight
the feathers are spread out, the lateral ones showing
pure white from beneath the blue-gray of the others.

Many other birds were seeking food in the shrubs
and trees close by, but I doubt if any succeeded in
finding as much as did the tiny gnat-catchers. They
were content with small fry, seemingly believing that
" many a mickle makes a muckle." No insect of any
size escaped their gaze. Gnats, mosquitoes, moths and
flies were spied out a dozen, yes fifty, feet away, and
with one straight dive and a click of the bill, the days
of the insect were ended forever. The birds seldom
missed their aim although in one instance one flew
full seventy feet and caught a flying form too small
for me to see at that distance. The insects preyed
upon must have been poor in nutrition or else the
gnat-catchers are veritable gourmands, for while I
watched them each one caught, on an average, three
to the minute, which would be 1,800 for a day of ten
hours.

The usual alarm note, similar to that of the cat-bird
but much softer, was not heard during the hour that
the birds were observed. At intervals one would
utter a faint chirp or chuckle, as if talking to another.
A low *pit-ut-ut-e* sound was also occasionally made.
According to Coues, the gnat-catcher spends days in
such incessant activity as that which I noted, "till
other impulses are stimulated with the warmth of the
advancing season, and the sharp accents of the voice
are modulated into sweet and tender song, so low as
to be inaudible at any considerable distance, yet so

faultlessly executed and so well sustained that the tiny
musician may claim no mean rank in the feathered
choir."

Each pair, after mating, seek some tree with a grayish bark, usually an oak, maple or apple, and finding
a horizontal limb or convenient fork, they begin their
nest, building it principally from hair and the fine
fibres of various plants which they weave very closely
and compactly together. Finally they cover the whole
with a coat of lichens, fastening them on with the
finest of wool or the silk of spiders' webs. This
lichen covering serves the useful purpose of a mask,
rendering the color of the nest almost exactly that of
the bark of the tree on which it is built; thus hiding
it from the keen eye of the young oölogist walking
beneath, or the keener eye of the crow or hawk flying
above. But there is one eye sharp enough to detect
it. For no matter how deep and dark the ravine in
which a nest is hidden away; no matter what aid of
nature has been called into use in rendering it inconspicuous to the view of other animals, necessity seems
to lend a preternatural sharpness to the vision of the
female cow-bird, enabling her to discover, whenever
needed, a safe place of deposit for an egg, destined to
become at no distant day an orphan which will be a
heavy burden to its foster parents.

The nest of the blue-gray gnat-catcher when completed, is usually very small, and is cylindrical in form,
not hemispherical, like that of most other birds. One
which contained five eggs, taken on the 2nd of last
May, was but 5⅜ inches in circumference by 2¾ inches
in length, and weighed only 3.7 grams. But the cow-

bird cares nothing for the size or form of the chosen
asylum for her young. If it is only large enough for
one egg, it is sufficient for her wants, and she forth-
with appropriates it to her use without even a " by
your leave" to the rightful owners. And so, very
often, among four or five delicate little gnat-catchers,
there is found a large chuffy youngster, whose demand
for food is incessant, and if supplied in sufficient
quantity, he will in a day or two fill the entire nest,

Fig. 20—Cow-bird.

and smother beneath his greater bulk the lives of the
rightful occupants. It is one of those numerous cases
of a struggle for existence in which the most over-
bearing, ugliest and strongest survives, instead of the
fittest.

However, I suppose that the modern evolutionist
would say, that in this case ugliness and brute strength
are necessary qualities of the " fittest," and that nature
has ordained that the cow-birds shall increase in num-

bers as the millionaires of to-day grow in wealth, only
at the expense of their weaker brethren.

But one—or rather two—cow-birds' eggs laid last
season did not hatch, and it was of them that I started
to write. On the 22nd of April, while out for a walk,
I discovered a pair of gnat-catchers building about
thirty feet from the ground, in a maple tree. A week
later, on passing near the spot, I saw that the nest
had assumed massive proportions for one of that
species, and on climbing up to investigate, found that
it contained a single cow-bird's egg. The owners,
however, had not deserted it, for they soon appeared,
circling rapidly around, and uttering their shrill cries
of distress. I left them immediately, merely suppos-
ing that they were young birds, not fully up to the
times in nest building, and therefore had formed a
large, loosely-constructed nest, instead of a small compact one, as is usually the case.

On the 5th of May I again visited the tree, and
found that the birds had abandoned the nest without
laying in it, and were building a new one in the top
of a tall oak a short distance away. Removing the
old nest carefully, I carried it home in order to com-
pare more closely its size with the one taken a few
days before. On measuring it carefully I found its
circumference to be $9\frac{1}{2}$ inches; its length, $4\frac{7}{8}$ inches;
and its weight 12 grams, or about $3\frac{1}{2}$ times that of the
one first taken. Judge of my surprise when, on examining it thoroughly, I found that it was a double
nest, or rather a "two story" one. The lower part,
or first floor, was neatly and closely built, and in it
was found a *second cow-bird's egg.* It had evidently

been laid shortly before the nest had reached the
usual size of such structures, and the builders, on dis-
covering it, had immediately set to work and covered
it entirely over, and then built up the sides of their
house about $2\frac{1}{2}$ inches higher. This upper portion
was very loosely constructed, and had evidently been
built in a hurry to meet the exigencies of the case.
But, alas, for the expectations of our feathered friends!
No sooner had the second floor neared completion
than Mrs. Cow-bird paid them another visit, and left
behind her a reminder in the shape of a new egg.
This was too much for bird endurance. The gnat-
catchers deserted in disgust the home over which they
had spent so many anxious moments, and set to work
to build a new one, in which, let us hope, they reared
their little family unmolested by unwelcome guests.

MID-SUMMER ALONG THE OLD CANAL.

Shady groves, green grass, wild flowers, and the sweet songs of our native birds in their chosen haunts —who in the hot and dusty days of mid-summer does not dream of such delights? True, parks there are, Collett's and Forest, easy of access both, and each with its own peculiar charms, but for him who likes a degree of privacy the crowds which gather there with their bustle and noise savor too much of those to be seen daily on the streets of the city.

To many persons, and especially to any one interested in the objects and doings of nature, there comes at times an irresistible desire to leave, as far as possible, all signs of civilization; to plunge, as it were, into a wilderness and spend an hour, a day, or a week in solitude. To a resident of Terre Haute, one of the best and most accessible places for such a day's outing is along the old Wabash and Erie Canal between Conover's and the Five Mile Pond. True, it is not a wilderness, but there is many a shady nook and quiet, secluded spot where one may rest free from interruption and enjoy the pleasing odors of wild flowers and the songs of sweet singing birds. On a recent date, a day hot, sultry, and disagreeable in the impure air of the city, I sought this favorite resort of mine, and it is of the birds I saw and heard, the flowers I met with, and a few of the thoughts which

(82)

welled up and were snatched from oblivion by a ready
pencil, that the present article has to do.

By street railway to Collett's Park, and then westward by the gravel pit road, one reaches the canal at
one of its loveliest points near the southern edge of
Conover's Pond. At the gravel pit I stopped awhile
and saw the puny power of a single man gradually
undoing what the mighty glaciers of the " Great Ice
Age " had done long centuries ago. Pebbles of manifold kinds and sizes were being exposed once more
to the sunlight after being hidden from it for—how
many thousand years? The iron pick wielded by the
workman pulled each from among its fellows and disturbed the quiet which had reigned with them since
that former day when, after years of rolling, crush-
ing, onward movement, they had been dropped, by
the melting of a mighty bulk of ice, on the spot
where they had since lain. And now they must be
carried out to do duty for man ; to receive the crush-
ing effects of his wheels of iron, and, perhaps, by
them be crumbled into dust after having successfully
resisted the giant powers of the glaciers of long ago.

Ah, the grandeur of the work which has been done
by nature's forces in the past for the benefit of the
races of the present! The sunlight of the old Car-
boniferous age did a work which now turns the wheels
of industry throughout the world and the glaciers
brought from afar the materials for our roadways and
deposited them where they would be needed, yea, at
the thresholds of our very doors.

Beyond the gravel pit, where in June glistened the
waters of a broad, spreading pond, now gleamed in

the August sunshine the golden yellow rays of the
bur-marigold. Acre upon acre of them nodded to
me from afar, while at my feet, on the roadside, their
western cousin, the fetid marigold, made known its
presence, not so much by its rays, which are few and
small, as by its disagreeable ódor which is its most
significant sign.

Down into the field of marigolds I took my way,
desiring to meet them face to face and learn what they
had to say of the summer's haps and mishaps. As
seen from the brow of the hill all appeared to be marigolds, but when among them much of the yellow was
found to be due to another handsome Compositæ with
a homely name, the sneeze-weed. The land on which
they grow was formerly cultivated but of late years
has been overflowed in spring, the water standing on
the ground each season until June. The owner,
therefore, has turned the land over to these wild
plants, and how they revel in their freedom! What
a struggle among them for existence now that man's
hand is not among their enemies! Two species of
marigolds, one devoid of ray flowers, the other with
the showy golden-yellow rays; two of smartweed,
one cocklebur, the sneeze-weed, and the fog-fruit, the
last a handsome creeping member of the Verbena
family—all growing in this damp rich soil in such
luxuriance as to literally hide the surface of the earth
from view. Which will be successful at the close of
the struggle? Which, in five, ten, or fifty years, will
be master of the soil? Perchance a stranger from
some western plain or from one of Europe's vales will
then have come and by its properties of prolificness

and endurance have driven out these native plants and in the end become the victor of them all.

Much has been said and written about the beauty of the cardinal flower but not too much, for among all our wild plants which bloom from August to October it ranks without a peer for brilliancy of color and gracefulness of form. And so, when amidst the tall rank grasses near the margin of the pond I came suddenly upon several of them, their bright red pennons contrasted so vividly with the omnipresent yellow of the sneeze-weed and the marigold—their purity and beauty seemed so enhanced by their surroundings that I could but fall upon my knees and do them reverent homage.

Reaching at last the old tow-path of the canal I threw myself down in a shady bower and gave way to revery. The time was when the tandem mules by scores passed daily over the very spot where I now sat. Then, busy commerce reigned supreme and man, bowing to her imperious demands, carried by the produce of the world. Woolens and silks, lumber and iron, coffee and teas, drugs and spices, indeed all the varied articles needed by a young and growing commonwealth, went up and down this artificial road of water.

Now, how changed! Commerce no longer, but nature reigns supreme. The tow-path is covered with the saplings of elm, ash, red-bud, and sycamore of fifteen to twenty year's growth. Wild birds of many species surrounded me on every side. From the topmost twig of a stately elm a southern mocking-bird sang for me a delightful medley of mimicry. It seemed as though a dozen different birds joined in the song; the notes of the cat-bird, jay and thrasher, chewink, pewee and robin, being each easily recognized as components of the medley.

As the clouds banked up in the west and north a turtle dove cooed softly above my head. A rain crow in a neighboring oak uttered his harsh refrain about the rain that was sure to come. Bob-white, in a stubble field on the hill above, whistled at intervals his

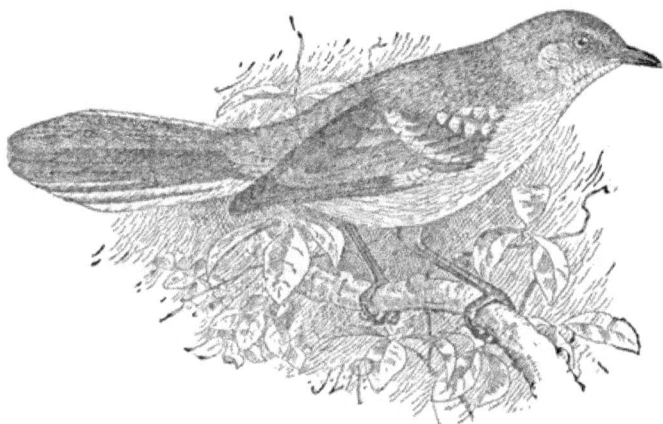

Fig. 21—Southern Mocking-bird. (After Judd.)

summer note. A yellow-breasted chat in a near-by thorn tree scolded incessantly, as only a chat can scold, at my intrusion on his domain; while, blithest of all, was the song of the indigo-bunting and the merry warble of the vireos which were heard on every side.

Such sounds as these were uncommon here forty years ago. The silence of the treeless tow-path was then seldom broken except by the mule boy's "gee, ga-lang there," or "git up, gol-dern ye."

But the canal became too slow for our advancing civilization. The iron horse took the place of the mule. The engineer in blue jacket and overalls with smoke begrimed face and oily hands, that of the mule boy. The ungainly canal boat with its snail-like pace has been succeeded by the "limited express," which follows not the winding course of a water pathway,

Fig. 22—Rain Crow or Yellow-billed Cuckoo. (After Beal.)

but dashes onward over hill and mountain, through valley and plain, on a smooth and even steel track; while the "gee, whoa-haw" of the canal boy has given place to the shrill toot of the locomotive.

Many seeds of many weeds and old-fashioned flowers were scattered along the tow-path in those old days. Now their descendants are seen in many places, flourishing and blooming more luxuriantly than if cultivated by the hand of man.

Ah, those "old fashioned" flowers, as we call them, how they bring up the memories of long ago! Of a country garden and door-yard where the "bouncing bets," "butter and eggs," "holly-hocks" and "lark-spurs" gave forth their beauty and their odors to entrance our childish mind, and, in the innocence of childhood, were thought to be the handsomest flowers that grew. To-day they are still attractive. Not by comparison with other and newer friends among the flowers which are far more beautiful than they; but because they ever recall the memories of yore when the struggle for our existence was borne by other hands and each day brought its round of pleasures unshadowed by any thought of the morrow.

Other plants there are in abundance along the old canal which man, in his ignorance, calls "homely weeds," ne'er seeing their smaller points of usefulness or beauty. Among these is the mullein with its long spikes of yellow flowers and thick velvety leaves. Those near its base, now withered and dry, last winter formed a beautiful rosette close to the ground and gave shelter and protection to many an insect both beneficial and injurious. Along waysides and the borders of barren fields the mullein has its favorite home. Ever an evidence of the presence of man, to-day it finds a congenial lurking place along the pathway of his former road of water.

Milkweeds, too, flourish there in great profusion and often reach a height of six feet or more. Their handsome umbels of purple flowers are very attractive but prove a veritable death trap to many a bee and unwary insect which visits them in search of honey. For the pollen of each flower, instead of being in numerous small grains, as is usual in other plants, is massed into a few waxy and adhesive bunches; and is so arranged that when an insect touches a certain point the pollen mass moves suddenly upward and clings by a slender stalk to the leg of the visitor.

If, as sometimes happens, a mass adheres to each of three or four of its legs the unhappy insect is so encumbered that it cannot move and so remains a prisoner until death. Almost every insect which leaves a bunch of milkweed flowers carries away one or more of the waxen masses, and as it goes immediately to an adjacent bunch, cross fertilization is thus more readily and surely accomplished than by ordinary methods of pollen distribution.

Queen of all our creeping or trailing shrubs is the trumpet creeper with its large pinnate leaves rivaling the emerald in their shade of green, and its giant trumpet-like flowers so attractive to humming bird and humble-bee. Mid-summer is the time, and the banks of the old canal the place, to see this creeper in all its primitive beauty. There the soil is congenial and bush and shrub furnish a ready support to which its aërial rootlets freely cling, thus forming many a snug retreat in which the nest of woodland songster is securely hidden.

Numerous other wild flowers, many of which are as deserving of praise as those above mentioned, are

now blooming along the old canal; for at this season
its banks

> " Are gay with golden rod,
>   There blooming grasses nod,
> And sunflowers small and yellow turn ever to the sun;
>   Quaint darky-heads are there,
>   And daisies wild and fair,
> In everybody's garden each flower's the loveliest one."

Space forbids the detailed mention of others. Go,
my reader, and see them for yourself. The flowers
and birds and bugs are there, and one has but to open
eye and ear to see and hear them. One warning,
however, before you go. The only way to thoroughly
enjoy and be benefited by any outing is to leave all
business cares and responsibilities behind. Think
not for a single moment of the time lost, the dollars
slipping away during the absence from the store, the
office or the shop. For the time being let "by-gones
be by-gones" and " to-comes be to-comes." Yield
only to the pleasures of the present. Dream only of
the duties of the day. Strive to gain rest, knowledge
and inspiration from the objects met with, and enjoy-
ment and benefits, not to be measured by dimes or
dollars, will be yours.

THE IRON-WEED.

Belonging to the great *Compositæ* family of plants are many of our worst weeds, such as the rag-weed, horse-weed, white-top, thistle, burdock, and last but not least, that scourge of our blue-grass pastures—the iron-weed. The name, *Compositæ*, is given to the family from the fact that its members have their small, yet perfect flowers densely crowded together into a head, which is enclosed by an involucre or cup formed of several circles of modified leaves called "bracts"; this involucre performing the same protective function for the compound mass that the calyx or outer green envelope does for the ordinary separate flowers of other families. The object of this massing together of a great number of small flowers into a large head is that they may more easily and certainly attract the attention of insects and thus secure their fertilization. Taken singly, the flowers are too small and inconspicuous to attract separate attention, but by huddling themselves together into a showy mass they have proven themselves very successful plants; so much so, indeed, that the family is now the largest known in the vegetable world.

Many of the *Compositæ*, as the sunflowers and asters, have the outer flowers of the head enlarged into the so-called "rays," thus increasing their showiness, while

those which have no rays, but which have all the
flowers of the head alike, are said to be "discoid."

Pre-eminent among the latter group, on account of
their size, abundance and hardiness, are the plants
known as iron-weeds—two species of which— *Vernonia fasciculata* Mich., and *Vernonia noreboracensis* (L.),

Fig. 23—Iron-weed. (After Britton.)

are found in Indiana.   However, the two are so nearly
alike that none but the practical botanist can distinguish them, and hence they will be spoken of in this
connection as one.

In the latter part of August, 1887, I spent a day in
a large woodland pasture in central Indiana.   The

effects of the noted drought of that season were then visible in all directions. Vegetation every-where was dying or dead. All nature had put on an unwonted garb whose prevailing color was brown. The grass of the meadow had been cured into hay before it was cut. The leaves of the maple and beech were shriveled and dying. No blue lobelias greeted, as in Augusts gone by, my wandering footsteps. No cardinal flowers waved their red pennons above the sedges of the swamp, for both swamp and sedges were things of the past. Only the coarse iron-weed with its cyme of purple flowers seemed to be flourishing in the parched, dry soil; holding its own where all else was perishing—thus proving itself well worthy its name— tough and indestructible as iron.

In the great contest for supremacy forever going on among all plants as well as among all animals, the rag-weed, fox-tail, white-top, etc., go down before the creeping, smothering power of the Kentucky blue-grass; but this rough, ungainly weed ne'er gives up the struggle, and in many places grows as rankly in the farmer's best lowland pastures as does its cousin, the greater rag-weed or horse-weed, along the margins of his cultivated bottom fields. It is seemingly becoming more abundant each year, and at present is undoubtedly the worst weed with which the stock farmers of Indiana have to contend. Let us note briefly some of the characters which render it so tenacious of life and so difficult of extermination.

Its perennial roots are stout and fibrous, and each autumn are filled with a sufficient supply of nourishment to give the stalk of the ensuing year a good start in life. They radiate in all directions from the base of the stem, spreading over an area of several square yards and penetrating the soil in search of moisture to such a depth as to render abortive any attempt of man to pull the plant up bodily, roots and all. In rich soil the stem grows to a height of six to eight feet and is leafy to the top. The leaves are linear to lanceolate-oblong in shape, sharply toothed, and so innutritious that none of the higher animals, not even the sheep, will feed upon them.

The only insect enemies of the plant, so far as noticed, are the black blister beetle which attacks the leaves when other food is scarce, and a small gall fly whose larvæ feed upon the juices of the flowering branches.

Many species of bumble-bees and butterflies visit its blossoms in search of nectar and pollen, and thus aid materially in their fertilization. The flowers in each head number, on an average, twenty-five, each of which produces a single seed. On one specimen of medium size were counted 743 heads, so that 18,575 seeds, each capable of becoming a fully developed iron-weed, were borne by that plant alone, and the majority produce as many, or more. To secure a broad dissemination each of these seeds bears at maturity a tuft of light brown bristles known as the pappus, and by its aid the seed may be wafted by the wind miles away from the parent plant. Again, as the iron-weed grows in greatest luxuriance in the lowland pastures near small streams, many of the seeds fall upon the water and are borne onward till they lodge against some bank or are buried in the sediment deposited by an overflow; places well suited for their future growth. In these ways the weed is continually spreading into pastures which heretofore have been entirely free from it. Taking into consideration that it is a native plant, and therefore well suited to our soil; the character of its roots; the immense number of seeds produced; the modes of their dissemination, and its almost total exemption from the attacks of injurious insects, it is no wonder that it is so well able, not only to hold its own in the struggle for existence, but also to increase in numbers from year to year.

As to the methods of extermination, many have been tried, but, too often, in vain. One which can be relied upon, but which in many cases is impracticable, and moreover is fatal to the pasture as well as to the iron-weed, is cultivation of the soil. Continued stirring among its roots the plant cannot withstand and so soon succumbs before the onslaught of the plow and the hoe. Many persons believe that cutting the weed during the flowering season will destroy it, but they have their labor for their pains, its roots being perennial, and the plant being, therefore, not dependent upon the seed to carry it over the winter. One farmer assured me that he had mowed the iron-weeds from his pasture every August for 17 years, and that they were still as abundant as ever. Another, after mowing for years, went to work and grubbed them out, but failed, of course, to remove all the pieces of roots from the ground and the crop of the next season was not appreciably diminished. A third has very nearly solved the problem of their extermination, and it is to his experience that I wish to call especial attention.

He has a pasture of 80 acres, and at present it takes
but two hours' mowing, twice or three times a year,
to rid it of the weed; whereas, just across the way is
a pasture where it holds undisputed sway from May
to August.

This third farmer had studied botany and knew
that all plants have their food assimilated, that is,
fitted for their use, in their leaves. The crude food
materials, carbon-dioxide, water, and various nitrates
and other salts in a soluble form, are gathered from
the air and the earth by the leaves and roots; but in
the leaves alone these materials undergo certain chemical changes and are transformed into starch and other
foods which can be used directly by the plant in its
growth. Any excess, not needed for growth, is stored
up in the seeds, buds, under-ground stems and roots,
for the purpose of furnishing the young plant or the
growing shoot sufficient nourishment to live upon
while it is developing leaves which in time will gather
and assimilate a new supply of food.

This farmer, knowing all these facts, reasoned that
if he cut his iron-weeds in May, again in June, and
still a third time in July, the perennial roots would
not be filled with nourishment at the end of the season,
as their extra supply would be needed to start new
shoots each time the old ones were destroyed. Moreover, the leaves would not have time sufficient to produce much excess of food to be stored up in the roots,
as all their powers would be taxed to furnish food for
the rapid growth necessary for the new shoot to arrive
at maturity before the season was over. He proceeded to act upon his conclusions with the above

mentioned result. The perennial roots of the old weeds weakened and in time died, and the only ones which he now has to keep down are those which each year spring from seed grown in other places. The work of mowing the weeds three or four times a year, at first a difficult one, gradually became less and his pasture was in time almost redeemed from the scourge which had rendered it comparatively worthless. On the other hand his neighbor, who lets the weeds grow until August and store up a good supply of nourishment in the roots, not only loses the summer's pasture but retains the weeds from year to year.

It has been said that all things in nature have their use—that nothing exists but for a purpose. It is the work of science to discover and make known the use of nature's objects; and day by day her secrets are gradually being exposed, thereby advancing man in civilization, by enabling him to better control the ravages of those existing forms which are injurious to his interests. If, however, the iron-weed has a use, other than that shown in the beauty of its flowers, no one has yet discovered it. But there is time; for of the thousands of plant forms which exist, we know the uses of only a few, as corn and hemp, ginseng and blood-root. Let us hope that some valuable medicinal or other property will soon be discovered in the iron-weed and a reason for its existence thereby pointed out to the doubting humanity of the present.

Meanwhile the naturalist will go on admiring the beauty of its bloom; for however coarse and repulsive the stem and leaves may appear, yet each head, with its 25 or 30 dainty florets so prettily grouped within

their protective cup, reveals a striking beauty to the
true lover of nature.   And when in the glamour of an
August morn he stands upon a hillside and views
acre upon acre of the broad purple cymes waving in
the valley beneath, all memories of the plant as a pernicious weed are blotted from his mind by the attrac-
tiveness of the scene before him.

WASHINGTON'S MONUMENT, MARENGO CAVE.
(See page 133.)

TEN INDIANA CAVES AND THE ANIMALS WHICH INHABIT THEM.

Caves are uncanny places. So thinks the average person, when, for the first time, he stands in the entrance of one of these under-ground cavities and glances along its walls of stone until his gaze reaches the point where the shadowy rays grow dim and eternal darkness begins. The chilly current of air which comes rushing forth, and the solemn silence of the depths beyond, beget a sensation of awe which it is difficult to suppress. Only a long experience in traversing subterranean passages will accustom a person to their surroundings and enable him to enter a cavern with anything like that degree of indifference with which he passes from daylight into the darkened rooms of a dwelling.

In the mind of the naturalist who is a tyro in cave exploration, there is mingled with this feeling of awe one of great expectancy. He hopes to add some new and wonderful facts to his store of knowledge. Another world is to be opened up before him. He is to travel through passageways worn in solid stone by the slow eroding power of water. He is to see in actual process of formation those pendent stalactites of creamy, crystalline calcite of which he has often read. He is to meet for the first time whole races of animals bereft of the sense of sight. To him, therefore, the under-ground journey promises much more than to the ordinary observer who makes it mainly for the novelty and the scenery which it affords.

The Sub-carboniferous Limestone area of southern Indiana contains many sink-holes and caves within its bounds. This area is, for the most part, embraced in the counties of Owen, Monroe, Lawrence, Washington, Orange, Harrison and Crawford. Going southward from the center of the State, the sink-holes first become a prominent feature of the surface in eastern Owen and western Morgan counties, and are found in numbers thereafter, in the area mentioned, until the Ohio River is reached, beyond which, in Kentucky, they are said to be still more numerous, in

Sink-holes. many portions of that State averaging 100 to the square mile. These sink-holes vary much in size, sometimes being but a rod or two across, and again embracing several acres in extent. They are, for the most part, inverted cones or funnel-shaped cavities, and, where small, usually have the sides covered with a matted growth of vines and shrubs. Where larger, trees of varying size are often found growing from the scanty soil on the sides or from the bottom of the sink. If one will examine closely the lowest point of a sink-hole, he will usually find a crevice or fissure through the limestone, or sometimes a large opening which, if it be possible to enter, will be found to lead to an under-ground cavity —a cave.

Both sink-holes and caves not only owe their origin, but usually their entire formation, to the slow, unceasing action of rain or carbonated water upon the limestone strata in which they occur.

Carbon dioxide is present everywhere in the atmosphere, constituting about three parts in 10,000 of the volume thereof. The condensed vapors, falling as rain, unite with a portion of this carbon dioxide, and form a weak carbonic acid or rain-water. Wherever this comes in contact with limestone, it brings about a chemical change. By this change the limestone is dissolved and carried onward with the seeping or flowing waters.

Formation of Sink-holes and Caves.

In the beginning of a sink-hole, the rain-water seeps through a crevice or joint of the limestone to a lower stratum, along the surface of which it finds a passage. By gradually dissolving the stone, this passage-way becomes increased, until finally a large cavity is formed immediately below the surface. The unsupported weight of the latter causes it to gradually sink downward and assume the inverted cone shape above mentioned. The opening at the bottom becomes larger, allows more water to enter, and a more rapid dissolving takes place between the layers. As soon as the under-ground passage has become large enough to allow a good-sized stream to enter, the process of erosion or abrasion is added to that of the solvent action of the water and the enlargement of the passage goes on much more rapidly. This gradual enlargement continues for hundreds, perhaps thousands, of years and results in a cave, varying in size according to its age, the amount of water flowing through it and the softness of the rock dissolved or eroded. The larger caves possess great vaulted rooms, deep pits, high water-falls and streams of water; some of the streams being large enough to allow the ready passage of a good-sized boat.

From the above it will be seen that sink-holes and caves are closely related, the latter, in fact, being largely dependent upon the funnel shape of the former to collect the surface waters and direct the flow thereof. A number of sink-holes often connect by narrow and tortuous channels with the same underground passage, the latter increasing in size with the addition of each new branch, until finally it attains majestic dimensions.

The rooms and passages of limestone caverns are often, after their formation, partially filled by those handsome forms of crystalline limestone, called stalactites and stalagmites. These are seldom, if ever, formed in great numbers, except where the passages or rooms are close to the surface. The water, charged with carbonic acid, filters slowly through the soil, and, entering the narrow crevices and joints between the layers of stone, seeps downward until it pierces the roof of an under-ground cavity. Here the slowly dripping water comes in contact with **Formation of** the air of the cave. The liquid is **Stalactites and** evaporated and the solid particles of **Stalagmites.** carbonate of lime, dissolved from the rocks with which it had come in contact, are left behind. Each successive drop thus deposits or leaves a solid particle, until finally a pendent cylindrical mass, called a *stalactite*, and resembling in general form an icicle, remains suspended from the roof.

Where the water, thus oozing through the roof of a cavern, is greater in quantity than can evaporate before it falls, it drops from the stalactite to the floo.· below. There it splashes outward' and in time evaporates, leaving the solid particles brought down. These accumulate one on top of another until finally a cylindrical or cone-shaped mass protrudes upward, slowly growing in size, each successive layer being distinct from the preceding. This upward rising mass is a *stalagmite*. It is almost always greater in diameter than the stalactite above it. Often the two, in time, meet and a column, or *stalacto-stalagmite*, of crystalline limestone results. Down the sides of this

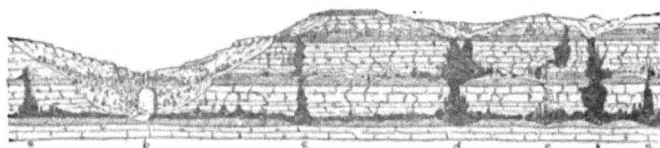

Fig. 24—Showing the formation of caverns in limestone.

s, Sink-hole; *a, c, d, f, g*, rooms in cavern; *b*, natural bridge, formed by the sinking of the roof of a former very large room; *e*, passages, showing numerous stalactites. (After Shaler.)

the incoming waters slowly flow instead of drop, evaporating and leaving their solid particles as they move, thus increasing in size the diameter of the column. If this action continues long enough, the whole passage or room may be filled by these deposits and all semblance of a cave obliterated.

It will thus be seen that water, where it flows freely and rapidly through massive beds of limestone, dissolves and erodes great cavities therein; where it seeps and oozes through such beds it tends to fill up the cavities already formed. Where the slowly flowing water has passed through large masses of pure crystalline limestone, the resulting stalactites and stalagmites are often very clear, almost translucent. Where sediment and mud is carried down with the carbonate of lime the resulting formations assume a dirty brown and unattractive appearance.

Many small caves, and doubtless some large ones, exist in southern Indiana whose presence is, as yet, unknown. In searching for them the bottom of a sink-hole will be the best starting point, the only thing necessary being to blast or dig out the cavity commonly found there, until it becomes large enough to allow a person to enter.

PORTER'S CAVE.

The mouth of Porter's Cave is in the north-eastern corner of Owen County, very close to the line between Owen and Morgan counties, and the source of the stream which flows therefrom is distant about one-half mile to the north-eastward. The cave is little more than a narrow, water-worn passage-way through the rocks, which at this point consist of St. Louis and Keokuk limestones, the former comprising the roof and the latter the floor.

The mouth of this cave is the most beautiful of any visited in the State. It is in the side of a hill at the head of a narrow canyon or gulch, which has been eroded by the stream which flows from the cavern. From the floor of the cave to the bottom of this gulch the distance is 33 feet, down which the stream flows in a perpendicular water-fall. The mouth of the cave is fifty feet wide and 14½ feet high, the roof extending out in a broadly arched front several feet beyond the the face of the water-fall below. The rock down which the water flows is covered with moss, and in the early morn, when the sunbeams light up the interior of the cave for a distance of 75 or more feet, and glisten and sparkle from the mossy background of the falling water, the scene is a most entrancing one.

The cave can be entered only by a narrow footpath on the northern side of the mouth. Twenty feet back from the entrance the roof becomes flat, and for almost 100 feet is comparatively smooth, being composed, apparently of one immense slab of limestone. In this distance the width gradually narrows to 30 feet. The floor is wholly of rock, in some places covered to a depth of several inches with sediment and loose stones brought down by the running stream. The latter, for the first 270 feet, is from four to eight feet wide and two to five inches deep. It meanders from side to side of the floor, making the frequent crossing of it a necessity. Beyond 270 feet it covers the entire floor to a depth of from six to twenty inches, and farther exploration must be made while wading.

The cave salamander, *Spelerpes maculicaudus* (Cope), inhabits this cave, several specimens being found within 200 feet of the entrance. They were clinging to the damp walls and showed little fear when approached. The raccoon, *Procyon lotor* (L.), visits the cave in numbers and evidently passes entirely through it, as was evinced by the tracks, which were very plentiful along the margins of the stream. To their visitation is probably due the absence of crayfish and other crustaceans, no specimens of which were noted. Four kinds of cave flies were found on the walls from seventy-five feet on back as far as exploration was made. A few cave myriapods dwelt beneath stones within 150 feet of the entrance, and three kinds of spiders were taken from the floor or ceiling. One of these, *Theridium porteri* Banks, had not before been described. It was found on the walls or roof of the room at the source of the cave, and near each specimen was often two and always one small globular cocoon, suspended by a single thread from the roof or a projection of the wall. Scattered threads of webs were also noted, but ran in no definite direction.

But few stalactites occur in Porter's Cave and they are dirty brown in color. At a point 250 feet from the entrance a very large one partially shuts off the passage-way; and 645 feet in, a similar one which has had its lower portion broken off, is found. At 750 feet the roof becomes so low that one has to stoop, and the width is reduced to 18 feet. From this point onward both height and width gradually diminish until at 852 feet it became necessary to crawl through water, and further exploration was abandoned. It is claimed that in a dry season persons have passed entirely through the passage, crawling for several hundred feet and then emerging into a low room near the source. A visit to the latter showed that it was not a true sink-hole, but a passage-way worn through the rocks in the side of a low hill. The opening was ten feet wide and about four feet high and a short distance back expanded to twenty-five feet in width, but soon narrowed again to eight feet, and 150 feet from the entrance the roof came down close to the water and stopped farther progress. Except to the naturalist there is little attraction about Porter's Cave other than its mouth; but that alone is well worthy a visit by all who enjoy the picturesque and beautiful in nature.

COON CAVE.

This cave is located in the south-western part of Monroe County, about eight miles from Bloomington, the county seat. The entrance is a perpendicular pit or well, forty-six feet deep and about six feet in diameter. The top of this pit is at the bottom of a rather shallow sink-hole and the descent into the cave was made by a rude ladder which had been constructed of poles by some previous explorer. At the bottom of the pit one finds himself on the edge of a passage-way, about ten feet high and nine feet wide, which extends both to the right and the left. The right hand passage is but about ninety feet long, the roof and floor gradually converging and being but a foot or so apart at that distance. Thirty-five feet from the entrance is a hole in the floor of this right hand passage, through which one can be lowered by a rope fifteen feet to the floor of a lower passage, twenty-five feet long, ten feet high and six feet wide, which extends nearly parallel to the passage above. By the side of a smaller opening is a stalactite, seven feet six inches long and five feet five inches in circumference, suspended from the bottom of the upper floor into the passage-way beneath.

The left hand passage comprises the greater portion of the cave. It varies in height from four to twenty feet, averaging about eight. But little stooping or crawling is necessary, but much climbing over rough stones and up and down steep, rugged slopes has to be done, the floor in most places being covered with great masses of fallen rock. Two hundred and forty feet from the entrance a crevice leads off through the walls on the right. By crawling along a ledge of projecting stone for about 100 feet, we reached the edge of an opening large enough to admit the body of a man, and by the aid of a sapling, bearing numerous short prongs or remains of limbs, which we found in place, we descended twenty-eight feet into a lower passage, about sixty feet long and ten feet wide. Here we found some shallow pools of water, but no living forms, and nothing in the way of scenery to reward us for the labor of getting down and up.

In numerous places the floor of the main passage has a deep cleft near its center or on one side, varying in depth from eight to twenty feet, and in width from a few inches to three feet and more. In several other places, notably 340 feet from the entrance, are openings or deep pit-holes, similar to those already mentioned, leading down into lower passages; the latter, however, of small extent. The main passage begins to narrow about 575 feet from the entrance, and 100 feet farther on is but three feet wide. At this point a branch turns to the left and leads downward into a lower room of small size. A short distance beyond this branch the main cave ends in a small crescent-shaped room, in the farther end of which, 750 feet from the entrance, is a deep crevice in the floor, filled with water of exceeding clearness. This pool of water was four feet wide and appeared but three or four feet deep, but actual measurement showed it to be nine feet, three inches in depth. The length of the pool could not be determined, but it extended down a branch passage to the right, covering all the floor thereof as far as one could see. For two or three feet above the water-line the walls of this room are covered with small but most beautiful crystals of calcite,

1. Right-hand passage.
2. Left-hand passage.
3. Cleft in floor.
4. Lower passage.
5. Blue pool.

MAP OF COON CAVE.
Monroe County, Ind.

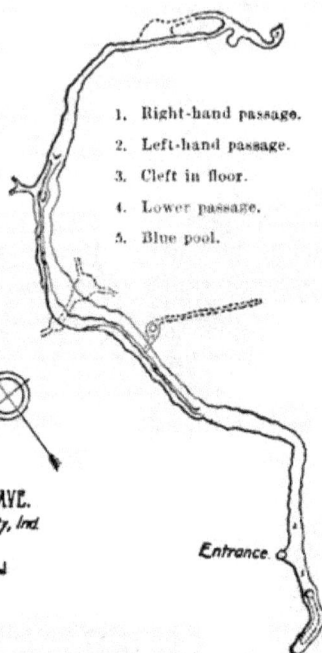

Entrance

Fig. 25.

which reflected the light of our candles in a most brilliant manner. Numerous small stalactites of the clearest crystal stud the walls and project from the crevices of the roof, while the floor is largely composed of calcite, derived from the overflow and subsequent evaporation of the water from the pool. This room is, in truth, a fairy grotto, decked with jewels resplendent, and a view of it will well repay for all the time and toil necessary to step within its bounds.

Animal life was represented by but few forms in Coon Cave. Several bats were found hanging, head downward, from the roof, but all were of the single common species, the little brown bat, *Vespertilio subulatus* Say In winter they are found in the cave by thousands, suspended from the roof and projections of the walls, their bodies remaining in a state of comparative torpor for months in succession.

Several specimens of a cave-inhabiting beetle, *Quedius spelæus* Horn, were found beneath stones within 300 feet of the entrance. This is a twilight form, the adults of which are not wholly blind, and usually lives in the dim light near the entrances of caves, feeding upon the excrement or decaying remains of such animals as frequent the place. A few flies were found on the walls and near them two kinds of spiders were taken—showing that here, as elsewhere, the fly is followed by the spider, with the ever ready invitation "to walk into his parlor."

By far the most common form of life in this, as in many other of the caves visited, was a small whitish insect, *Degeeria cavernarum* Pack., about one-eighth of an inch in length and belonging **The Cave** to the order *Thysanura*. Like most **Spring-tail.** other cave-inhabiting insects, it occurs only in comparatively moist places, often swarming by thousands beneath or on the surface of damp rocks, especially where organic matter, such as the remains of lunches, drippings of candles, decaying wood, etc., is scattered. It has the power of leaping several inches by means of a long, spring-like appendage bent under the hind body, which on being released throws the owner high in the air. The motion thus produced may be likened to that effected by a spring-board. These little acrobats, however, carry their spring-boards with them wherever they go and hence have come to be known by the common name of "spring-tails." The one under consideration doubtless forms much of the food of the small spiders, harvestmen and beetles which frequent the floors of the caves.

ELLER'S CAVE.

The entrance of this cave is at the bottom of a sink-hole, 100 feet in diameter, which is located in a woods about five miles south-west of Bloomington, Monroe County. The cave itself is a double floored one, the upper and older floor being dry, and the more recent and lower floor having a stream of water flowing through the greater part of its length.

The entrance, about six feet wide and six and a half high, descends gradually for about fifty feet, and there expands into a room twenty feet wide, thirty feet long and twenty-five feet high, which serves as a vestibule or starting point for both floors, the entrance to the upper one being in the wall, about eight feet above the floor of the common entrance.

Two passages lead from this vestibule to the lower floor, one to the right through a narrow winding cleft in the rock, and then down to the bed of a stream, along which, by crawling, one can advance until he

comes out into the second passage, fifty feet from its
starting point. From here onward for 210 feet the
lower passage leads through a water-worn crevice
from two to four feet wide and three to fifteen feet
high, the stream sometimes covering
its bottom, and again running in a
channel cut beneath one or the other
of the sides.

In this stream were found two species of small crustaceans. One was a

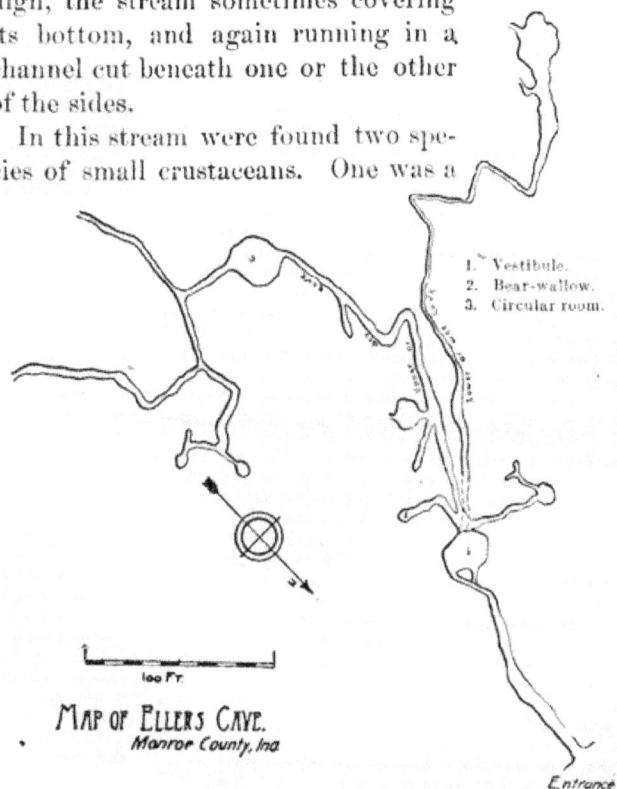

1. Vestibule.
2. Bear-wallow.
3. Circular room.

100 Fr.

MAP OF ELLERS CAVE.
Monroe County, Ind.

Entrance

Fig. 26.

shrimp, *Crangonyx gracilis* Smith, three-fourths of an
inch or less in length, which is often found in wells
and springs in central Indiana, and had probably been

washed into this and other caves in which it was found by the heavy rains of the season. The other and smaller species, *Cæcidotæa stygia* Packard, is a true subterranean form and was the most common crustacean noted in Indiana caves. It was usually found singly, swimming or crawling slowly through the water of small cave streams, and was easily picked up with a pair of forceps. Its body is flattish like that of a sow-bug, but is oblong and more slender, reaching a length of one-third of an inch. These crustaceans probably furnish much of the food for the blind fish and crayfish which often inhabit the same streams with them.

A Cave Crustacean.

Fig. 27—*Cæcidotæa stygia.*

Three hundred feet from the cave entrance the lower passage ends abruptly in a room fifty feet high and ten feet wide, the sides converging in an angle to form the roof. On the left, about twelve feet from the floor, is an arched opening, and through it comes a roaring sound of falling water. With difficulty one climbs a slippery bank and, passing through this opening, finds a most magnificent scene for so small a cave—a great cylindrical pit or shaft, twenty feet in diameter and sixty feet high, down which, on the farther side, falls a stream of water. A large bowl-shaped cavity, twelve feet deep, has been worn by the falling water in the limestone below the level at which the pit is entered. Descending into this, it was found that the stream flows out through a passage to the left too low for exploration.

8

Returning to the vestibule we climbed to the entrance of the upper floor, and, passing a short distance within it, found two passages diverging. One to the left, but forty feet in length, ends blindly against a bank of hard clay. Here had been, in days of yore, a bear-wallow and the marks of bruin's claws were numerous and plainly visible in the clayey walls. The right hand passage proved a long and tortuous one and had a number of short branches leading from it, one of which showed plainly the evidence of former inhabitancy by bears. This main upper passage is in most places seven to ten feet high, with a width of five to seven feet. Two hundred feet from the vestibule it became necessary to crawl for about thirty feet through a space one foot high by two feet wide, when we emerged into a circular room thirty feet in diameter by three and a half high, the floor of which contains a vast amount of bat guano. Beyond this the passage forks into three branches, each of which was explored as far as possible, the longer one reaching 400 feet from the vestibule before its small size barred further progress. The floor of this upper cave was covered in many places with a yellow ochery clay. In this, in several places, were found some handsome acicular crystals of selenite. No water was found on the upper floor, except at the farther end of the galleries, where it stood in shallow pools. These were evidently quite near the outer surface, as the shells of several land snails were found near by the water.

SHILOH CAVE.

The entrance to this cave is at the bottom of a sink-hole a few rods north of Shiloh Church and about seven miles north-west of Bedford, Lawrence County. Except after a heavy rain, no water flows through the entrance, but a stream runs the entire length of the main cave, entering it from beneath a great mass of fallen rock which has partially closed the entrance, and meandering from side to side on the floor in its onward course. On entering, one descends rapidly for about twenty feet, and then reaches the general level of the main passage. This passage is from fifteen to twenty-five feet high and about the same width for 2,000 feet, which was as far as it was explored, the water becoming too deep to wade beyond that point. It far exceeded any of the previous caves visited in the number and size of its stalactites and stalagmites, many of which were of exceeding clearness. In the words of Prof. John Collett, who visited the cave in 1873: "The lofty sides are draped and festooned with stalactites, sometimes hanging in graceful folds, or ribbed with giant corrugations. Above, the roof and overhanging sides bristle with quill-like tubes, fragile as glass, each tipped with a drop of water which sparkles in the lamplight like a crystal jewel."

Three hundred feet from the entrance three jets of water pour down from the right wall of the cave and add to the size of the stream along its floor. These falls vary in height from seven to ten feet, and together they produce a roaring sound which is echoed far along the main passage-way.

From this point onward the walls are dripping more or less and are fringed with small stalactites. About 900 feet from the entrance are two large stalagmites, one of which, named by Collett "The Image of the Manitou," has been broken. Originally it must have been six feet in height and eighteen inches in diameter.

In a pool of the stream in the main passage were secured two of the small aquatic insects known as "water boatmen." They belong to the order *Hemiptera*, and to the genus *Corisa*, and were the only "true bugs" taken in Indiana caves. They were probably accidental visitors, since their compound eyes were fairly well developed.

In the same pool were numerous specimens of the blind crayfish, *Cambarus pellucidus* (Tellkampf). This curious crustacean was found in a number of other Indiana caves, and probably inhabits every one in which there is a permanent water supply. Careful examination of cave bed streams ought, also, to show its occasional occurrence outside of its subterranean homes.

The Blind Crayfish.

During heavy rain-falls the water rushes with great violence through the caves and doubtless often carries the crayfish out to the rivers. Here its light color, soft shell and defenseless condition would prove such a heavy handicap that in the struggle for existence its life would be of very short duration. It is usually found in shallow pools with muddy bottom rather than in rapidly flowing water. It moves slowly with its antennæ spread out before it, and gently waving to and fro, feeling, as it were, every inch of its way. It is wholly non-sensitive to light, and seemingly so to sound, but when disturbed by any movement in the water it is extremely active; much more so than ordinary terrestrial forms, leaping upward and backward with quick, powerful, downward blows of its abdomen.

Several branches leave the main passage of Shiloh Cave, but all but one are short in length. The one exception turns to the right about 1,500 feet from the entrance and extends in a south-westerly direction. At

Fig. 28—Blind Crayfish.
(Three-fourths natural size.)

first it is a high, narrow fissure with the jutting walls bearing many stalactites. A stream of water covers the entire floor and from far in the distance comes a murmuring sound caused by a succession of water-falls, four in number and in size small, which occur at short intervals along the passage. Wading through pools, clinging to corners of jutting ledges,

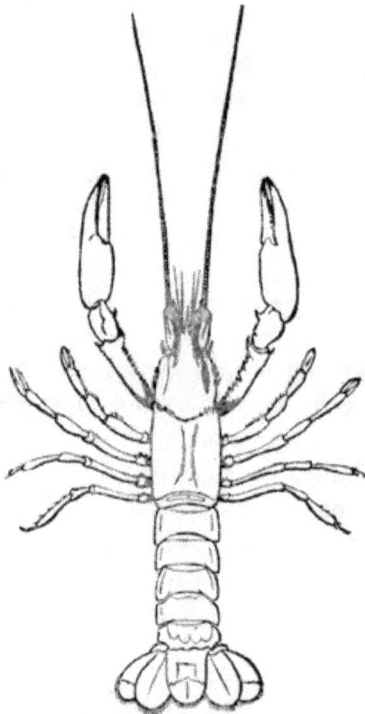

climbing over slippery, perpendicular banks we made
our way until finally the passage began to rise, and
the limestone gave way to a dark shale and this in
time to a light colored clay. We were 900 feet from
the fork and thought we were nearing the surface and
would soon find our way above ground, when all at
once our lights went out and we staggered backward
through utter darkness, escaping, as if by a miracle,
the clutches of the deadly choke-damp which lurks for
unwary explorers amidst the deepest recesses of this
cave.

Beyond the point where the right branch leaves it,
the main passage continues in a southerly direction
and was explored until the back water from the dam
at the mouth of the cave became too deep to wade.
While preparing to leave the cave a heavy thunder
shower came up and the water soon poured in torrents
through the sink-hole and adding its volume to that
of the enlarged stream within the cave, quickly covered the entire floor to a depth of nearly two feet.

DONNEHUE'S CAVE.

The mouth of this cave is located near the foot of
one of the bluffs of White River, 500 yards distant
from that stream and two and one-half miles south-
west of Bedford, Lawrence County. From the mouth
of the cave a small stream finds its way, the source of
which is in a sink-hole three-fourths of a mile distant
in a north-easterly direction. The stream is greatly
enlarged after a heavy rain and by its erosive action
the cave is constantly but slowly increasing in size.

Entering the cave, one finds himself in a commodious room, 10 feet high and 48 feet in width, the floor of rock, covered in places to a depth of two or three feet with alluvial drift. Fifty feet back this narrows to 12 feet in width and a short side passage puts off to the left, in which a number of the cave salamanders, *Spelerpes maculicaudus* (Cope), were found. This handsome batrachian was taken in a number of the caves visited and doubtless occurs in all Indiana caverns which contain streams of water or damp rooms near the entrance. In life it is a **The Cave Salamander.** bright orange-yellow with very numerous black spots, which, on the back and sides, vary much in size and shape. The body is quite slender and reaches a total length of 6¼ inches.

It is usually found clinging to the walls within 150 feet of the entrance of the caves, especially in crevices and crannies just above flowing streams or pools, but never in the water. While its eyes appear as large and normal as those of allied terrestrial species, its sense of sight seems to be limited. It remains quiet when discovered and shows little fear until touched, when it scrambles deeper into a crevice or beneath some fallen rock on the floor. Even when a candle is put within a few inches of its head it does not move until it feels the heat. Its food probably consists of such insects and small crustaceans as are found along the margins of the streams.

Back 180 feet from the mouth, the main passage of the cave is 6½ feet high by 6 feet broad, the stream on the floor being about three feet in width and three inches deep. Farther on this stream deepens and several pools were found in which the water was two or more feet in depth. At a distance of 325 feet the passage forks, and from the right-hand branch came so strong a current of air that it was impossible to use candles and lanterns had to be substituted. The change in lights made, the right hand passage was found to be a narrow, winding one, about 150 feet in length, and to lead back into the main passage about 100 feet farther from the mouth than the point from which it started. All these branches are through the solid rock and are only water channels three or four feet high and about as wide.

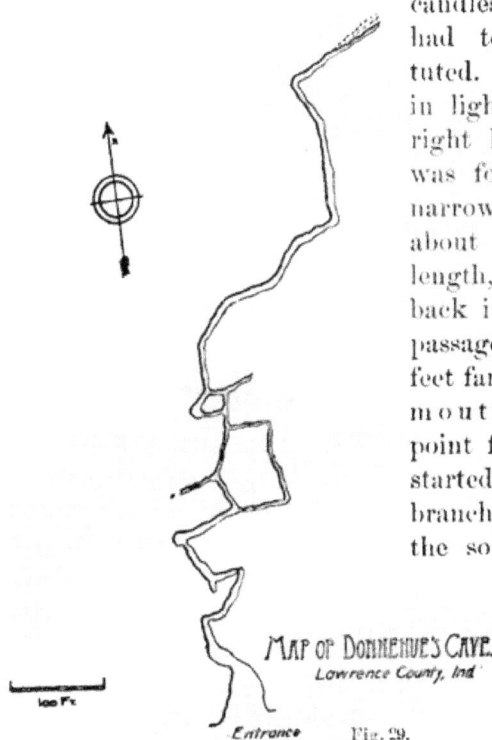

MAP OF DONNEHUE'S CAVE.
Lawrence County, Ind.

Entrance   Fig. 29.

100 Ft

Beyond 425 feet, several side branches were found to contain water too deep to wade, or to soon become too low for further progress; in fact, the rock is more honeycombed with small passages than in any cave visited. The main passage, however, at about 500

IX.

Mouth of Shawnee Cave.

feet from the mouth, enlarges to a height of 40 feet and a width of eight to ten. This portion was, for the most part, dry, the stream having disappeared in one of the low channels already mentioned. In some places two floors are found; in others the greater part of the upper floor has fallen in, leaving a portion in the form of a natural bridge spanning the passage from side to side. At a point 950 feet from the mouth the upper passage ends against a perpendicular wall of rock, from near the top of which is a passage onward, but too high from where we stood to admit of entrance. The lower passage was followed to about the same point, where it became two feet high and three feet wide and almost filled with water, thus barring further progress. But few stalactites were found in the cave, and they were mostly of small size and unattractive appearance.

SHAWNEE CAVE.

Among Indiana caverns the mouth of Shawnee Cave ranks next to that of Porter's in picturesque beauty. Indeed, by some it is classed as more attractive. The mouth of the cave is found at the head of a deep gorge worn through the limestone by a good sized stream which flows from the cave and down the gorge to the broader valley beyond. Many centuries ago the cave extended the full length of the gorge, and the waters of the stream flowed directly from its mouth into the valley. The roof of the under-ground channel finally became so thin that it collapsed, the gorge was then started and, as the centuries went by, grew in length, the cave ever becoming shorter by the continued falling of the roof. Both gorge and cave are located about three miles south-east of Mitchell, Lawrence county, in a region noted for the beauty of its scenery.

Three passages open directly into the mouth of the cave. The right hand passage has the level of its floor about five feet above that of the entrance, while the opening on the left is 12 feet above the bed of the stream and very difficult to enter without a ladder. The middle passage extends straight back from the common vestibule or main entry. The latter is twenty-five feet long, twenty-one feet high and eighteen feet wide, but at its farther end is reduced to the narrow middle passage between great masses of limestone. The water in this passage is waist deep and explorations must be made by wading or in a light canoe. One hundred feet within is a magnificent cascade, where the stream rushes and leaps down a narrow passage with such violence that the noise is plainly heard at the entrance.

The right hand passage, for the first 100 feet, is about ten feet high by fifteen wide, with a clay bottom and a roof on a level with that of the vestibule. It then expands into a large room, 230 feet long and forty feet wide, which lies east and west at right angles to the entering passage. This narrows at the west end to twenty feet and at one point the outer air flows in through a small opening in the roof. From near the smaller end of the room a narrow passage starts off to the southward and can be traveled for 200 feet, when it becomes too narrow for further advance. Along this passage a small stream flows, disappearing through a hole in the floor near the entrance to the larger room. Other than this, both right and left passages leaving the main entry are dry.

The passage at the left of the main entrance to the cave is about 150 feet long by twenty broad, and contains no points of especial interest. No stalactites worthy of notice are found in this cave. The name "Shawnee" has been given it from its being near the center of the former hunting grounds of the Shawnee Indians. It was doubtless used by them as a place of shelter since many relics have, from time to time, been found about its mouth. In the early settlement of this region the nitrous earth on the floor of the two dry passages was used in the making of saltpetre; and the stream flowing from the main cave was afterwards dammed and utilized for a number of years in driving a woolen, grist and saw mill.

This stream is one of the largest issuing from an Indiana cave.. It flows for a long distance underground and in several places south of Shawnee Cave the roof of its subterranean passage has caved in, causing deep ravines at the bottom of which the stream meanders, until it reaches a point where the roof of stone remains intact, and the entrance of a new cave begins, into which the waters disappear, as

> "Alph, the sacred river, ran
> Through caverns measureless to man,
> Down to a sunless sea."

In this stream the blind fish, *Amblyopsis spelæus* De-Kay, occurs in numbers, though never more than two

and seldom but one are seen at a time. When full
grown this fish reaches a length of four and a half

The Blind Fish. inches. The body is colorless, the
scales very small and the young are
born alive. No external traces of eyes are to be found
in adult specimens, but the loss of sight is in part
compensated by numerous touch papillæ, arranged
in ridges on the sides and front of the head.

These eyeless fishes move very slowly through the
water, usually near the surface and close to the edges
of the deeper pools which they inhabit. They are
wholly non-sensitive to light, but extremely sensitive

Fig. 30—Blind Fish.
(Three-fourths natural size.)

to touch or any jar or motion of the water. They
were readily caught by putting a dip-net very gently
into the water a foot or two from them and then
making a quick forward and upward scoop. If in
still, deep water, they seem to glide, or rather float,
on and on, propelled by a scarcely perceptible motion
of the caudal fin. One must think of them as ever
surrounded by an intense darkness, the prey of every
fish-loving animal, as mink or coon, that can swim
and see in the darkness, the white skin of the fish
readily revealing its presence if the least gleam of
light reflects from its sides.

Concerning the sense of hearing as developed in this species Prof. E. D. Cope has written as follows: " If these Amblyopses be not alarmed they come to the surface to feed and swim in full sight, like white, aquatic ghosts. They are then easily taken by the hand or net, *if perfect silence is preserved*, for they are unconscious of the presence of an enemy except through the medium of hearing. This sense is, however, evidently very acute, for at any noise they turn suddenly downward and hide beneath stones, etc., on the bottom."

My observations do not bear out the above statement. I talked and even hallooed close to the fish without causing them to take alarm, but the least movement of the water frightened them, and they darted rapidly away, usually at right angles to the course they were pursuing. The sense of touch, rather than that of hearing, is, in my opinion, the one which has been intensified by long residence in the dark and silent recesses of the caves. In a number of instances, as the dip-net was raised quickly upward, the fish leaped several inches above the surface of the water in a vain endeavor to escape.

In one place where a stream flows out of a cave and through a deep ravine for about 200 yards, and then enters another cave, the blind fish were captured in both caverns within 100 feet of the openings, and there is little doubt but that they make their way through the open stream from one to the other. The caves and under-ground streams of southern Indiana doubtless form a more or less complete system of sub-terranean drainage, and through this the blind fish finds its way wherever the water is deep enough to allow it passage.

In captivity this fish eats very little. Dr. Sloan, of New Albany, who kept specimens in an aquarium for 20 months, says: "They have taken no food, except what has grown up in the water and on the sand in their tank. . . . Some of them would strike eagerly at any small body thrown in the water near them, rarely missed it, and in a very short time ejected it from their mouths with considerable force. I often tried to feed them with bits of meat and fresh worms, but they retained nothing. On one occasion I missed a small one and found his tail projecting from the mouth of a larger one; I captured and released him."

In nature they doubtless feed upon one another and upon the blind crayfish and smaller crustaceans which inhabit with them the streams of caves. A number of those captured were "nosing," as they slowly swam, the rocks along the sides of the pools, and it is possible that they gather some organic matter from the slime on these rocks.

CLIFTY CAVES.

The mouths of the two Clifty caves are about 200 yards apart, and are located at the head of a deep and narrow valley about three miles north of Campbellsburg, Washington County. Clifty Creek has its source in the streams which emerge from the caves, and flows in a north-westerly direction about four miles to White River, into which it empties. Its valley, especially the upper half, is noted for the wild and rugged scenery and the vicinity of the caves is a noted resort for pleasure seekers.

The caves are designated, respectively, by the terms "wet" and "dry," the former being the smaller of the two. Across the mouth of the Wet Cave a dam has been built, and the water emerges from it with sufficient force to turn the machinery of a distillery and grist mill; both abandoned, however, since their owner died, a few years ago. The mouth of the cave is a perfect archway in the solid limestone, fourteen feet wide and eleven feet from roof to bottom. The water behind the dam was two and a half feet in depth, and deepened rapidly as one went back, and the cave was explorable only by means of a boat, which was not at my command.

Dr. John Sloan of New Albany, Indiana, at one time went up the stream in the Wet Cave for about 200 yards on a raft of timber, at which point rapids were encountered, over which it was impossible to lift the raft, and the water above being too deep to wade, he was obliged to return.

The Dry Cave was explored for a distance of 2,650 feet, beyond which it was impossible to proceed. The entrance is larger than that of the Wet Cave, being eighteen feet high and twenty feet wide. Back 100 feet it narrows to thirteen feet in width, and, fifty feet farther, to about eight feet, the water at this point covering the entire floor to a depth of six inches. For the first 500 feet the main passage is very crooked, but beyond that point it is comparatively straight and extends in a general south-westerly direction. Like Porter's Cave, it is a mere water-worn passage, with no large rooms, few stalactites, and, in general, may be said to be monotonous. The stream on the floor winds from side to side of the cave, thus making the frequent crossing of it necessary.

In a shallow pool of water, 1,200 feet from the entrance of the cave, a fine specimen of the blind crayfish was secured, and about fifty feet distant, in a deeper pool of the main stream, numbers of a common species of an above-ground crayfish, *Cambarus bartonii* (Fab.), were found. Whether these seeing forms pass their entire lives in the total darkness of the cave or whether they make an occasional visit to the outside is a question as yet unsolved. The same species was found in several other caves and seems to have a liking for clear, cold water and underground resorts. If these habits be continued a " new species" of blind crayfish will, in time, result; for there is little doubt but that the ordinary eyeless form has evolved from a seeing one which, ages ago, found its way, voluntarily or otherwise, into the under-ground streams. Finding there the struggle for existence less deadly than among its numerous kin of the surface waters it slowly adapted itself to its surroundings. Having no need for its eyes they, in time, became aborted, for nature always rids her objects of every organ which, from a change of environment or habit, becomes to them useless.

Several specimens of above-ground beetles were taken from the margin of the stream in this cave, but they had probably been washed in by the heavy rains of the week before. The most interesting insect secured was a cave harvestman, *Scotolemon flavescens*

(Cope), which is closely allied to the common " daddy long-legs." It is, however, much smaller, and pale yellow or reddish in color. It frequents the surface of damp rocks and probably feeds upon the little cave "spring-tail" which was everywhere abundant.

Several short side branches diverge from the main one, and at a distance of 1,300 feet from the mouth a large branch turns off to the right, which was explored for about 400 feet, but not to the end. The main passage continues to the left, and at 1,800 feet a large rock

Fig. 31—Cave Harvestman. (Much enlarged.)

30x15 feet was found which had fallen from the roof and partially blocked the way. Two thousand feet from the entrance the passage widens into a room 100 feet across and four feet in height, which contains much fallen rock, but nothing else of especial interest. Beyond this the cave narrows again and varies from twenty to thirty feet in width as far as explored.

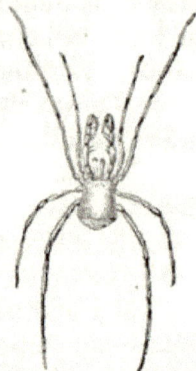

MARENGO CAVE.

This cave, which next to Wyandotte is the most noted in Indiana, is located a short distance north-east of Marengo, Crawford County. It has been known only since 1883, and the owners of the land on which the entrance is located were wise enough to prevent the ruthless destruction of the stalagmites and stalactites which form the main beauty of the cavern. Some

9

children playing about a sink-hole in September of that year, noted an opening which had been formed near its bottom by a recent falling of earth and rock, and venturing in, found the room now known as "Grand Entrance Hall." Afraid to go farther, they made known their discovery to other persons, and in a few weeks the entire cave had been explored. A building was soon afterward erected above the mouth, and stairways built, so that entrance into the cavern could be easily and safely made.

Thousands of visitors have since passed through this cave, and no one who is at all in sympathy with nature can come forth from its corridors and passages without feeling fully repaid for his peep into one of her under-ground chemical workshops. There, the only materials necessary are water and limestone. Given these, and time unlimited, the varied character and wonderful beauty of the products possible can only be realized by those who have spent a few hours in a cavern like Marengo.

Descending the stairways, after having been provided with a lantern and guide, the visitor finds himself fifty feet below the surface in the large vestibule known as the Grand Entrance Hall. This is a room fifty feet wide and twenty to thirty feet in height, the floor of dry earth, and with two passages diverging, one ascending to the right and leading through the Short Route and Crystal Palace, the other descending to the left and leading through the Long Route.

Taking first the latter, we found the main passage to be 12 feet high and about 20 feet wide. Scattered at intervals along its walls and roof were many stalac-

tites, some in groups, others singly, and all possessed
of fanciful names given them by former visitors or by
the proprietors and guides of the cave. One hundred
feet from the foot of the entrance is a slab of limestone, fallen from the roof, whose dimensions are
18 x 8 x 4 feet. This is known as "Fallen Rock," and
beyond it a short distance is, on the right, a passage-
way known as the "Cut Off," which leads to the
Crystal Palace. Continuing, the main passage widens
to 30 or more feet, and for a distance of 80 feet is
known as "Statue Hall." In this are some note-
worthy formations, the prettiest of which is "Mt.
Vesuvius," a large, rounded stalagmite. Above it is
a group of slender stalactites, down which a stream
of water trickles and gives a muddy character to the
floor for a distance of several hundred feet.

Crawling over the damp rocks were several specimens of a small, light yellow spider, *Nesticus carteri*
Em., which were quickly consigned to a vial of alcohol. It belongs to the group of "wandering spiders"
whose members spin no webs, and its food is doubtless,
the little cave "spring-tail" which occurred by thousands in the same damp area. Here, also, were taken
the first specimens of a blind myriapod, afterwards
found in great numbers in Wyandotte Cave.

Congress Hall succeeds Statue Hall and contains
along the edge of the ceiling some handsome formations, known as the "Giant's Mitten," "Mammoth
Pen," etc. From this hall the bed of an old stream
leads to the right beneath the massive limestone walls.
"Mammoth Hall," with a width of sixty-five feet and
a length of about 300, comes next in order, and con-

tains the " Elephant's Head," " Folded Lambrequin,"
" Bridal Curtains" and other fantastic formations of
carbonate of lime,
wrought in dark-
ness in the ages
past.

Beyond Mam-
moth Hall the
passage divides
and passes around
a mass of unerod-
ed limestone. The
branch on the
right rises ten or
fifteen feet above
the level of the
main floor and en-
larges into " Elks'
Hall," a room 190
feet long and
twenty feet high.
Beyond this hall
the two branches
soon unite and at
a distance of 1,000
feet from the en-
trance enlarge in-
to " Music Hall,"
a large room con-

1. Grand Entrance
   · Hall.
2. Cut Off.
3. Congress Hall.
4. Mammoth Hall.
5. Elks' Hall.
6. Music Hall.
7. Cave Hill Cemetery.
8. Creeping Ave.
9. Junction Room.
10. Fairy Palace.
11. Prison Cell.
12. Prison Bars.
13. Washington Ave.
14. Lovers' Retreat.

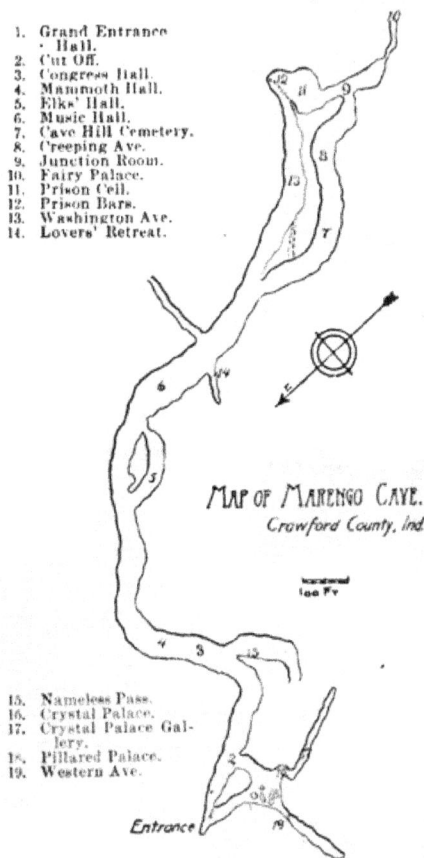

MAP OF MARENGO CAVE.
Crawford County, Ind.

15. Nameless Pass.
16. Crystal Palace.
17. Crystal Palace Gal-
    lery.
18. Pillared Palace.
19. Western Ave.

Entrance

Fig. 32.

taining a raised platform of rock, known as the
" Band Stand." A short distance farther on, a branch
goes off to the left which has been explored only by

X.

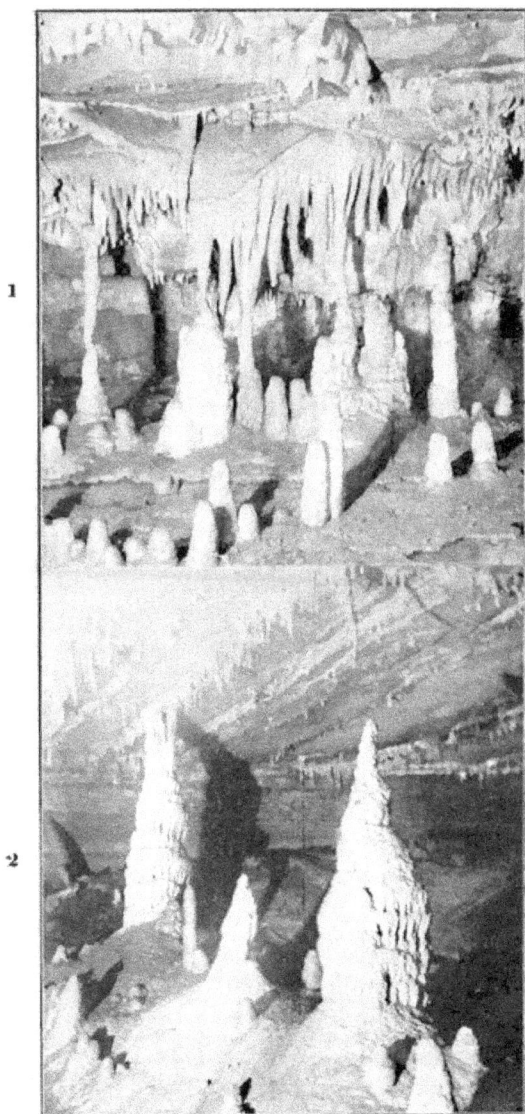

1. CORNER OF CAVE HILL CEMETERY, MARENGO CAVE.
2. TOWER OF BABEL, MARENGO CAVE.

guides, the ceiling being low and the scenery possessing no especial interest.

Fourteen hundred feet from the entrance the main passage again forks, the right branch containing "Cave Hill Cemetery." Herein are found some beautiful stalagmites and pillars, one of which, called "Washington's Monument," is among the most striking objects of the cave. Its height is four feet, eleven inches, and a foot above the base it is two feet in circumference. Composed of the clearest of crystalline limestone, it stands with its white surface gleaming in the dim lantern light, inspiring the visitor with a feeling of wonder as to how an object of such beauty and purity could have been formed in these depths of Cimmerian darkness. Another monument of greater size, but less imposing on account of its yellowish-brown color, is the "Tower of Babel"—ten feet high and six feet, eight inches in circumference. It stands among numerous smaller stalagmites, a short distance beyond Washington's Monument.

Beyond the tower of Babel the roof of the right branch lowers, and it is necessary to crawl through a narrow opening and then creep or stoop for quite a distance through "Creeping Avenue," passing meanwhile among many pillars, stalagmites and stalactites, varied in form and beautiful to look upon. Emerging from this avenue we stood erect in the "Junction Room" 2,000 feet from the cave entrance, and at the point where the branch which turned to the left at the entrance to the Cave Hill Cemetery meets the right branch through which we have traveled. Beyond this point the cave narrows and the roof comes down

within a foot of the floor. By creeping, crawling and
twisting from side to side we managed to get up a
slippery hill and through a small opening into the
" Fairy Palace," a place visited by few on account of
the difficulty of the way. Here we found the farthest
explored part of the cave, and in the small room, perhaps ten feet wide and five feet high, were thousands
of formations, which reflected the light in a most brilliant manner.

Retracing the way to the Junction Room, we turned
to the right into the " Prison Cell," a large room which
contains some of the principal features of the cave.
Here is the " Leaning Tower of Pisa," a stalagmite six
feet high, with the top inclined several inches beyond
the base; "Solomon's Temple," a group of slender
pillars six and a half feet in height and arranged in a
circle; "Administration Building," a pillar made up
of a series of circular layers of crystalline limestone,
piled one on top of another so as to cause the whole
to resemble a Japanese pagoda; " Bunker Hill Monument," formed on a fallen slab, story on story as the
preceding, besides many others as handsome, yet too
numerous to mention. We passed from the Prison
Cell, between the " Prison Bars"—a series of slender
columns six feet long and six to eight inches in circumference—into " Washington Avenue," the left
branch of the main passage at the fork near Cave Hill
Cemetery. This avenue is 450 feet in length, from
twenty-five to forty in width, and for about one-third
of its length the ceiling is so low as to require a stooping position in passing through. On the way are
many small stalagmites grouped in a straggling

fashion, "Grant's Army" and "Coxey's Army" being the names given to two of the largest assemblages. The floor of Washington Avenue is dry and for the most part composed of earth, with here and there a slab of fallen rock. Near the fork it descends for about fifteen feet and we enter once more the main passage, already described, and make our way along it to the "Cut Off" leading to Crystal Palace, passing on our left the "Lover's Retreat," a winding cleft which extends about seventy-five feet back into the solid limestone.

The Crystal Palace is the crowning glory of Marengo Cave. It is a small alcove or side room, ninety feet long, fifteen feet wide and about twenty-five feet in height. At the south end is a perpendicular wall along which is a drapery or vast sheet of stalactites, and from a projecting shelf are many slender stalagmites, the whole so grouped as to resemble a giant pipe-organ. The side walls are studded with hundreds of small and large formations, while from the roof hang, pendent, myriads of slender stalactites of the clearest crystal, which reflect with sparkling brilliancy the rays of the calcium or magnesium flash lights. By ascending a stairway fifteen feet, one finds himself on a balcony in the very midst of these formations and can pass back into "Crystal Palace Gallery," a low passage, about 150 feet in length, the floor of which resembles a relief map, being thrown up in many places in narrow corrugations and ridges, with here and there a pool of limpid water occupying the irregular and shallow depressions.

Descending the stairway and passing to the left,

we enter the "Western Avenue or Short Route," the
principal feature of which is the "Pillared Palace,"
where giant pillars, stalactites and stalagmites are so
numerous that it is with difficulty the visitor winds
his way between and around them. This portion of
the cave extends but about 150 feet in a westerly direction, and into it has been dug an artificial opening
from the surface, forty feet above. Retracing our
steps for the last time, we turn to the left at the
mouth of the Pillared Palace and pass through
another bower of beauty, the "Queen's Palace," a
small room whose walls are composed wholly of pillars and stalagmites. Beyond this we pass the "Diamond Dome," the largest stalagmite of the cave, thir-
ty-one feet in diameter and reaching from ceiling to
floor, and emerge into Grand Entrance Hall, which
was our starting point.

In the crevices of the walls of this hall the cave
salamander has its abiding places and among the debris which has fallen from the roof a number of white-
footed mice, *Calomys americanus* (Kerr), have taken
up their home. They have been noted
A Cave ever since the cave was discovered, but
Inhabiting seem to keep close to the entrance,
Mouse. through which, however, no light
passes. In the winter season they are very destructive
to sweet potatoes and pumpkins stored in the cave, and
at other seasons evidently subsist upon the tallow drippings and other refuse matter left by the visitors.
They differ much in appearance from above-ground
examples of the same species, having larger external
ears, longer whiskers and more protruding eyes.

Such changes have been brought about by their life in the darkness, and are but the beginning of a series of adaptations to their changed surroundings which will eventually result in a "new species of blind mice," provided such a life be continued for any length of time.

Careful measurements showed the total length of Marengo Cave, including the side branches, to be 3,850 feet. Within this distance of less than three-fourths of a mile are probably crowded more beautiful formations of crystalline limestone than in any other known cave of similar size in the United States.

Lacking the length, the lofty vaulted rooms and the grand scenery of Wyandotte, Marengo far excels that cave in the abundance and beauty of its stalactites, stalagmites and other like formations. To those who wish but a glimpse of under-ground life, we most heartily commend it, believing that a visit of a few hours will repay all who take an interest in the mysterious and beautiful in nature.

WYANDOTTE CAVE.

Next to Mammoth Cave, Kentucky, Wyandotte is the largest cavern in the United States. Its enormous under-ground halls and vaulted domes, its gigantic fluted columns and vast piles of fallen rock, are unexcelled in any other American cavern. Its situation among the rugged hills which form the breaks of the Ohio River, in a country as yet primitive in character, where game is plentiful, and fishing in the clear waters

of Blue River exceptionally good, make it a most
inviting spot for a summer's outing.

Around the hotel, situated close to the cave on a
commanding eminence in a natural wooded grove,
grow numerous forms of plant life which are strangers to central and northern Indiana, while in the cave
dwell many sightless animals whose habits of life are
yet unknown; so that the botanist and zoölogist may
add to the study of the cavern itself the pursuit of
their favorite subjects.

The first published account of Wyandotte Cave was
probably in 1833, in "Flint's Geography of the Mississippi Valley," as follows: "Like Alabama and
Tennessee, Indiana abounds with subterranean wonders in the form of caves. Many have been explored
and some of them have been described. One of them
is extensively known in the western country by the
name of 'the Epsom Salts Cave.'

"It is not very far from Jeffersonville. When first
discovered the salts were represented as being some
inches deep on the floor. The interior of this cave
possesses the usual domes and chambers of extensive
caverns, through which the visitant gropes a distance
of a mile and a quarter to the 'pillar,' which is a splen-
did column, fifteen feet in diameter and twenty-five
feet high, regularly reeded from top to bottom. Near
it are smaller pillars of the same appearance.

"The salt in question is sometimes found in lumps
varying from one to ten pounds. The floors and
walls are covered with it in the form of a frost, which,
when removed, is speedily reproduced. The earth
yields from four to twenty pounds to the bushel, and

the product is said to be of the best quality. Nitre is also found in the cave in great abundance, and sulphate of lime or plaster of paris."

The mouth of Wyandotte Cave is located in Jennings Township, Crawford County. The nearest railway, the "Air Line," passes through Milltown, which is nine miles distant from the cave, over an exceedingly rough road. From Corydon, the county seat of Harrison County, the distance is about twelve miles, and the road a fair one for southern Indiana. This route is a most pleasant drive in the summer or autumn, and leads one down the romantic valley of Blue River. For several miles the road follows along the limestone bluff on the right side of that stream, in many places having been excavated in the side of the bluff forty or more feet from the water below. From Leavenworth on the Ohio River, the nearest point for steamers, the distance is five miles.

According to measurements made by Prof. Collett, the Cave Hotel is 220 feet above Blue River, across whose narrow valley "Greenbrier Mountain, with sharp, conical peak and steep faces, belted with massive rings of rock and variegated with evergreen cedars, affords a scene of quiet, stately beauty." From the hotel a pathway leads down a gradual slope to the mouth of the cave 100 yards away.

That portion of Wyandotte known previous to 1850 is at present called the "Old Cave," and will be first described, since one must traverse a portion of it in order to reach the entry of the "New Cave," discovered in the year mentioned.

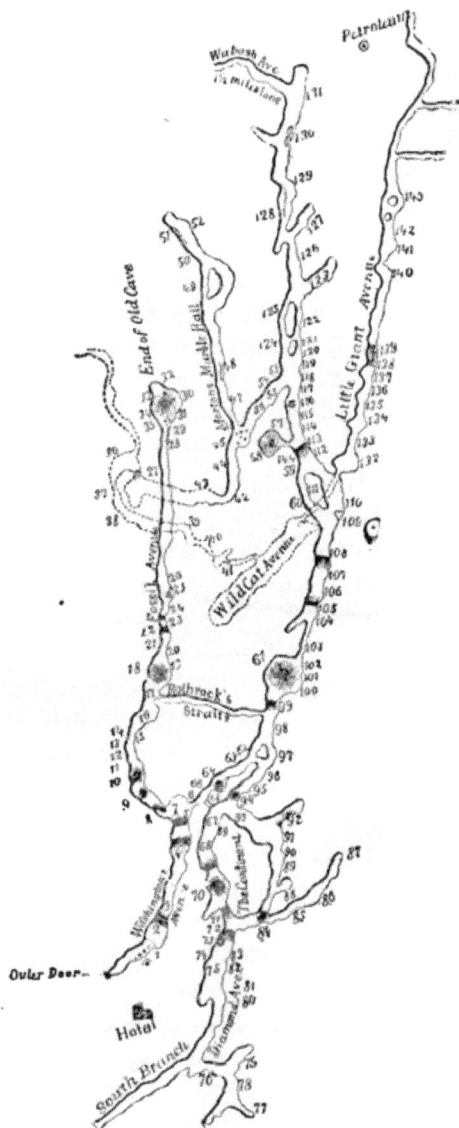

Fig. 33—Map of Wyandotte Cave.

47. Snowy Cliffs.
48. Indian Footprints.
49. Beauty's Bower.
50. Queen Mab's Marble Garden.
51. Fairy Palace.
52. Wyandotte Potatoes—Pebbles.
53. The Arm Chair.
54. Lovers' Retreat.
55. Ewing Hall.
56. Frost King's Palace.
57. Bowlder Flints.
58. Milroy's Temple.
59. Penelope Grotto.
60. Ulysses' Straits.
61. Rothrock's Cathedral.
62. Coons' Council Chamber.
63. The Rotunda.
64. Rugged Mountain.
65. Cut Off.
66. Counterfeiters' Trench.
67. Starry Hall.
68. Wyandotte Grand Council Chamber.
69. The Card Table.
70. Hall of Representatives.
71. Hill of Science.
72. The Alligator.
73. The Mound.
74. The Throne.
75. General Scott's Reception Room.
76. Ante-room.
77. Hovey Point.
78. The Pit and Sieve.
79. The Amphitheater.
80. Rocky Hill.
81. Muddy Fork.
82. Lost Rivulet.
83. Frozen Cascades or Curtains.
84. The Hippopotamus.
85. Fairy Grotto.
86. Neptune's Retreat.
87. Hermit Cell.
88. The Sepulchre.
89. Purgatory.
90. Calliope Bower.
91. Palace of the Genii.
92. Pillared Palace.
93. Creeping Avenue.
94. Junction Room.
95. Drawing Room.
96. Dining Room.
97. Delta Island.
98. Sandy Plain, 300 feet long.
99. Hill of Difficulty.
100. Monument Mountain.
101. Sulphur Spring.
102. The Auger Hole.
103. Lilliputian Hall.
104. Spade's Grotto.
105. Slippery Hill.
106. Hall of Ruins.
107. White Cloud Room.
108. Sentinel Office.
109. Bishop's Rostrum or Pulpit.
110. Journal Office.
111. Calypso's, or Island No. 2.
112. Cœrulean Vault.
113. Rugged Pass.
114. Chapel.
115. Vestry.
116. Josephine's Arcade.
117. The Parsonage.
118. The Junction.
119. The Lone Chamber or Ball Room.
120. Dry Branch.
121. Island of Confusion, or No. 3.
122. Grand View Island, or No. 4.
123. Sandy Branch and Air Torrent.
124. Newhall's Forum.
125. Grosvenor's Avenue.
126. Gothic Chapel.
127. The Gallery.
128. Indian Footprints.
129. The Den.
130. Ship in the Stocks.
131. Crawfish Spring.
132. Maggie's Grotto.
133. Joseph's Pit.
134. Lama's Bower.
135. Marble Rivulet.
136. Marble Hall.
137. Miller's Reach.
138. Andrew's Retreat.
139. Rode Rock No. 2.
140. The Devil's Elbow.
141. The Pit.
142. Langsdale's Basin.
143. Wash. Rothrock's Island.
144. Bourbonoi.

THE OLD CAVE.

The entrance of Wyandotte is twenty feet wide and six feet high; the roof arched, the floor of earth, with here and there a fallen slab of rock. For perhaps 100 feet we descended gradually and entered a spacious corridor known as "Faneuil Hall," forty feet wide, eighteen feet high, and probably 180 feet in length. Across the farther end of this hall a stone wall has been built and a doorway constructed, and through this one passes into "Twilight Hall," where the last rays of daylight disappear and the King of Darkness begins his reign. Here we stopped a few moments to accustom our eyesight to the changed conditions. Passing onward we soon entered the "Columbian Arch," an almost perfect semi-cylindrical tunnel, seventy-five feet in length. From this we emerged into "Washington Avenue," a grand passage-way, 275 feet long, thirty feet wide and forty feet high. Near the farther end is "Falling Rock," a huge mass of limestone, resting partly on edge, 33x16x14½ feet in dimensions, and weighing, therefore, about 535 tons. Ages ago it fell from the roof and assumed its present position; one which earthquakes have failed to change, but which appears dangerous to the average visitor who passes beneath its towering form.

Within Washington Avenue a peculiar pungent odor became noticeable, and inquiry as to its source brought information from the guide that in 1884 certain gentlemen from Evansville attempted to corner the onion industry of southern Indiana by buying up all the onion sets produced that season. Wishing a

XI

MOUTH OF WYANDOTTE CAVE.

suitable storehouse they rented room in the cave and deposited therein several hundred barrels of the sets. But, however suitable the pure cave air is for the preservation of sweet potatoes and other mild edibles, it failed to act in like manner on the onions, and they soon began to sprout and grow. All were lost and were allowed to remain in the cave, their shriveled skins and pungent odor still reminding the visitor thereto of an attempted financial "corner," which failed to materialize. Another odor, more strong and disagreeable, especially in autumn and winter, was noted at this point or before. It was that of the exhalations of thousands of bats which make the cave a winter abiding place. Their faint squealing notes and flutter of wings were the only sounds that greeted us from the depths of darkness beyond.

Passing under the falling rock and up a short declivity, we found ourselves in "Banditti Hall," fifty feet wide, forty to fifty high, and partially filled with rugged fallen stone, grouped in great masses on either side of the pathway. Stepping from slab to slab, we picked our way, until finally the guide called a halt, and lighting some "red fire," directed our attention to two outline figures formed on the ceiling above, by the scaling of the dark exterior from the whiter limestone. To one the name "Wyandotte Chief" was given many years ago by a correspondent of the *Cincinnati Times*, who wrote of it as follows: "We look up and see above the Falling Rock a mass of white limestone resembling the front of an Indian chief, with crown shorn to the scalp lock and fanciful earrings dangling from the ears. There he hangs, seem-

ingly suspended, like Mahomet's coffin, keeping his
dark and weary vigils, waiting to gloat over the death
of some daring pale-face, crushed by the Falling Rock
below." Upon the other figure, which resembles the
facial characters sometimes seen in Punch and Judy
shows, the fanciful name of "Betsy and I are Out"
has been bestowed.

Banditti Hall is the closing portion of the common
entry to both the Old and New caves. At its farther
end the opening leading to the Old Cave is seen on
the left, some twenty feet above the level of the floor,
while about the same distance below, on our right,
opens the doorway into "Fat Man's Misery," and the
New Cave beyond.

Climbing a steep ascent into the Old Cave, we found
ourselves at first in a passage-way ten feet wide and
seven feet high, with the floor of ochery clay a number of feet thick, the walls of oölitic limestone, and
the roof with here and there the more soluble portions
dissolved until it resembles a coarse-celled honeycomb
in appearance. Passing onward beneath "Pigmy
Dome," we entered the "Continued Arch," a long
passage-way, eight feet in height, ten feet wide, and
with an occasional crystal of selenite glistening on the
dry and dusty floor. From this we passed into the
"Canopy," a circular room, twenty feet in diameter
and ten feet high, with a smooth white roof. This is
succeeded by another long, low passage, where stooping is necessary for some distance, and then we passed
down through a narrow passage into "Lucifer's
Gorge," forty feet deep, with precipitous, jagged rocks
overhanging the sides. Up we climbed once more,

from rock to rock, and, reaching another opening, crawled over a natural bridge, and on hands and knees crept for seventy-five feet through the "Grecian Bend." Finally we emerged into "Odd Fellows' Hall," one of the grand under-ground rooms for which Wyandotte is noted This we measured carefully and found to be ninety feet wide, 210 feet long and sixty-five feet or more in height. The massive ledges of limestone forming the walls project toward the top, each layer a few inches farther than the one below, so that the ceiling is oval in shape, much narrower than the floor and appears to be hollowed out by successive fallings of rock. Great masses of fallen rock partially fill the room, and tens of thousands of the little brown bat, *Vespertilio subulatus* Say, hung head downward from the ceiling. We extinguished the lights, and their low squealing notes became instantly hushed; the only sound which broke the death-like stillness being a continuous faint and lisping noise, like the ripple of water over a distant water-fall, due probably to the rustle of the wings of such as were flying through the Plutonian darkness.

The bats choose as a resting place that part of the roof where small portions have begun to flake, giving a certain degree of roughness, or small crevices, to which they can cling. They cannot attach their claws **Cave Bats.** to a smooth surface, hence from large portions of the roof of a room they may be entirely absent. In places where they find a suitable foothold they congregate so closely together that it is difficult to pull one from the midst of a group. On a low ceiling in Salt Petre Cave, near

10

Wyandotte, an area one foot wide by one and seven-tenths feet long was measured, and the bats thereon were pulled off, one by one, and counted. Their number was 401 on the one and seven-tenths square feet. When pulled or knocked loose from the roof they fell to the floor, where they lay motionless for some time, and then began to flutter and crawl about, finally becoming lively enough to fly and find a new resting place.

Fig. 34 –Little Brown Bat.

Their squealing note was uttered only as we passed along, the light from the candles evidently disturbing those which had not yet entered their winter torpid state. Two other sounds they seem capable of making, one, the cry of a single bat in rapid, broken notes, resembling the song of a robin in a minor key: the other, a noise somewhat similar to the short, quick alarm note made by the common ground squirrel, *Tamias striatus* (L.), when it espies some intruder on its domain.

They show a remarkable sense of direction in their flight, passing, in a darkness so profound that man can see absolutely nothing, swiftly and unerringly through openings but a foot or two in diameter, without hitting the walls. The direction of flight seems to be, however, one of instinct or training rather than of reason, since when a door was first put in an opening in the cave through which they had been wont to pass in numbers, they flew blindly against it and were killed by thousands.

At Wyandotte, as elsewhere, the bats pass in numbers into the deepest recesses, being found abundantly in the "Senate Chamber" and sparingly near "Crawfish Spring," two miles or more from the entrance.

As is well known, bats are crepuscular in habit. They spend the day in darkness and the night in search of food. Such habits have, in the course of ages, rendered their eyes exceedingly small, their external ears large, their flight, like that of the owls and whip-poor-wills, noiseless. Several questions of interest, which to my mind are unanswered, arise regarding the cave life of these animals:

First.—In a cave where the temperature is 54° F. the year round, how do they determine when warm weather has begun out of doors?

Second.—How do those which spend the days of the summer season in the cave determine the approach of dusk?

Third.—How do they distinguish, in the intense darkness, those portions of the roof which are smooth from those which are rough enough to furnish a foothold?

On the right side and about fifty feet from the entrance to Odd Fellows' Hall is a pit-hole or perpendicular cleft in the floor, through which an average sized man can just squeeze himself. This is the opening into "Rothrock's Straits," a deep and narrow passage-way which connects with "Rothrock's Cathedral," a room of the New Cave.

From Odd Fellows' Hall we climbed by a rugged stairway and passed onward through narrow passages, and beside pits and chasms—the way ever seeming to grow rougher—the hills and valleys following each other in rapid succession. In one place we descended fully fifty feet and from the bottom noted on our right the perpendicular walls of rock known as the "Cliffs." Over these in ages past a drapery of stalactites has been thrown in graceful folds, resembling a cascade which in mid-air has been congealed into stone, and is most worthy of its name—"The Falls of Minnehaha." Below these overhanging cliffs is the gaping mouth of the "Pit"—a deep cavity leading by one drop fifty feet into space—as yet unexplored. From the foot of the Cliffs we made our way with difficulty up "Uncle Sam's Stairway" and then under the "Dead Fall," a large flat rock which lies at an incline across the passage, the upper edge supported by less than three inches of a thin rock projecting from the wall. From this point onward, for a distance of perhaps 1,000 feet, the way is a succession of steep climbs and steeper descents, varied by an occasional crawl on hands and knees and a final twisting of the body into shapes innumerable in order to effect the passage of the "Screw Hole," which forms the portal to the "Senate Chamber," the final room of the Old Cave.

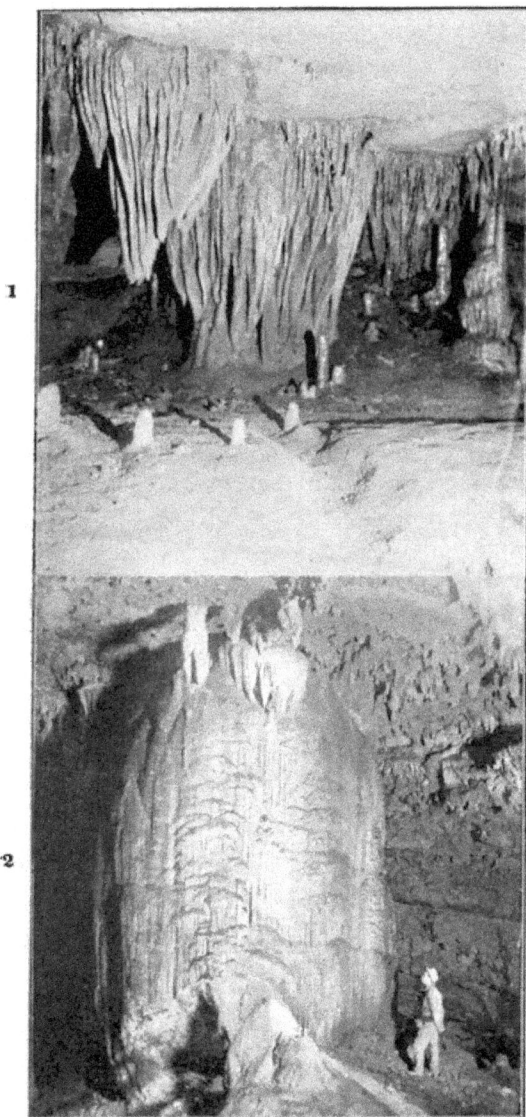

XII

1. TOBACCO SHEDS, MARENGO CAVE.
2. PILLAR OF THE CONSTITUTION, WYANDOTTE CAVE.
The opening at base of Pillar is a part of the Ancient Quarry.

Collett described the Senate Chamber "as a vast elliptical amphitheater, estimated at six hundred feet long and one hundred and fifty feet wide. The sides are built up with massive ledges of limestone, thinning and converging upward into a monster dome, with a flat elliptical crown fifty by twenty feet in diameter. The center of this vast room is piled up with a great mass of rocky debris fallen from the immense cavity above."

Other than the dimensions, this was an excellent description. Exact measurements show the room to be 144 feet long and 56 feet in width. The mass of fallen rock in the center, known as "Capitol Hill," is about thirty-two feet in height and is crowned to a depth of several feet with an immense mass of stalagmitic material. From the center of this mass rises from the top of the hill the grandest natural wonder in Wyandotte Cave—the great fluted column of crystalline carbonate of lime, known as the "Pillar of the Constitution." Perfectly cylindrical, seventy-one feet in circumference, and extending from the crest of the hill to the ceiling above, this enormous column exceeds in magnitude any similar formation in any known cave on earth. From the point where it first became visible in the dim light of our candles it appeared "like an immense spectral iceberg looming up before us, looking as if it had just arisen from the foaming waves of the ocean on a dark and foggy night." The entire column is composed of "satin-spar"—a rather soft, white, striated mineral, the purest form of carbonate of lime. From one side, near the base of the column, has been removed by the

Indians, or some prehistoric race in ages past, hundreds of cubic feet of this material.

Up to 1877 it was generally supposed that the whites had made this excavation. In 1864 J. P. Stelle wrote of it as follows : "For fifty years the people of a civilized—aye, a Christian nation, have visited the Senate Chamber, not as admirers of the great God who has reared for himself such a magnificent temple, but as vandals. All the most interesting formations within their reach have been broken up or carried away ; and even the great pillar itself has not been exempt from their attack, for an excavation has been made in its side which must have required days of hard labor, and from which large quantities of the purest white stone have been taken and scattered over the floor of the cave."

Prof. Collett, in 1877, found three glacial bowlders in the Senate Chamber, which, "from indications, such as wear and bruises, had been used as hammers or grinding pestles, and proved conclusively that that part of the Old Cave had been visited, if not occupied, by men of the Stone Age."

Rev. H. C. Hovey, in 1882, first claimed that the excavation had been made by Indians "more than 1,000 years ago," and that the "round or oblong bowlders" of granite rock were the implements with which the ancient quarrymen wrought, being used "in breaking from this alabaster quarry blocks of a portable size and convenient shape."

H. C. Mercer, in 1894, visited the quarry and mentions the finding, by Mr. Rothrock, of a pick made of stag's antlers and states that "the proof of Indian work at the spot was satisfactory and of a character never noticed and studied before the discovery of the site."

These constituted the recorded observations of the quarry up to the time of my first visit in July, 1896. It was then noted that the quantity of spalls and flakes of the material thrown over the side of the hill was very great, and that no digging had been done to discover the nature or thickness of the debris

An Ancient Cave Quarry.

on top of the hill, nor to more fully verify the statement that the work had been done by Indians. My time being limited, no excavations were made at this visit, but on a subsequent one, in November, 1896, I secured the services of a workman and shovels and again visited the place. Careful measurements showed that above the debris a space eight feet long, six feet high and five feet wide, or 240 cubic feet, had been quarried from the column. The top of the hill on which the column rests was found to be covered with an area 14 feet square of the debris, and through this, close alongside the base of the column, a trench was dug, eight feet long, three feet wide, and to the solid stalagmite beneath. It averaged four feet three inches in depth—i. e., at that point the debris or pieces of quarried material and other matter was that thick. A perpendicular section through this trench disclosed the following layers:

1. Ashes in a compressed, damp bed, with occasional flakes of stalagmite intermingled............... 14.0 inches.
2. Charcoal...................................... 1.0 inch.
3. Ashes, with flakes of rock .... 3.5 inches.

4. Rectangular flakes of stalagmite or satin-spar,
   varying in size from an inch or two square to
   pieces 8x3x1 inches, or even larger, with occasional traces of charcoal intermingled .... 28.0 inches.
5. Charcoal... 0.5 inch.
6. Flakes of stalagmite............................. 4.0 inches.

Six quartzose bowlders, weighing from three to six
pounds, were found scattered through the mass which
we threw aside, two of them within a few inches of
the bottom. They were worn with use, and on the surface of two or three of them were depressions which
appeared to be finger marks due to excessive use.
At any rate, they must have enabled the workman to
retain the rock hammer more firmly in his grasp.

The remains of horns of five different deer, which
mostly crumbled when disinterred, and numerous
small bones, also too much decayed to identify to
what animals they formerly belonged, were found at
intervals in the trench.

By digging in a few other spots it was found that
an area 14x14 feet, on top of this hill and at the base
of the column, was covered to an average depth of
three and one-half feet with the particles of stone
quarried. In addition to this, no less than twenty
tons of the material had been pitched over the hill.
Much, if not all, of this additional space was formerly
occupied by stalagmitic material, the base of the column flaring outward on this side, and when the space
already mentioned as having been quarried *above* the
debris is taken into consideration, there is little doubt
but that more than 1,000 cubic feet of the stalagmite
has been broken loose.

In October, 1898, I visited the quarry for the third time, and dug in the debris for several hours. Eight additional bowlders of quartzose and many pieces of horn, crumbling bone and baked clay, were brought to light. Five wedge-shaped pieces of rock, one of flint, the others of limestone, were also found among the flakes of stalagmite. These had irregular notches in their edges showing that they, together with the horns found in the debris, were most probably used as wedges to pry loose the pieces of satin-spar after the latter had been cracked by the stone hammers. Such horns and wedges of stone have been found in a number of caves in Europe, where, ages ago, they were put to similar use.

A large quantity of wood must have been necessary to have produced such a bed of ashes as was found. The carrying this in over the seventeen rough hills and through narrow passes, through which one has to crawl and where more than a candle is a burden to the ordinary visitor, must have entailed a vast amount of labor and leads one to suppose that the material sought was used for a purpose deemed especially valuable. What that purpose was I have not yet been able to ascertain, there being few objects made of stalagmite among the "Indian relics" in any collection or museum in the United States.

Down the sides of the Pillar of the Constitution tiny streams of water are constantly trickling, and, spreading out upon the top of the hill, quickly evaporate, leaving behind their solid particles to make thicker the crust of so-called "alabaster" which covers the rough edges of the mass of rocks. This action

will continue for thousands of years, until ultimately,
by continued accretions, this hill will reach the ceiling
and enclose entirely the wondrous pillar with its
flutings and carvings, wrought in ages past by that
magic graver—water.

Over the damp stalagmite forming the crest of
Capitol Hill numerous specimens of the cave myriapod,
Pseudotremia cavernarum Cope, were rapidly crawling.
Here also the little red harvestman, first noted in
Clifty Cave, was found in numbers, and is probably
utilized by the myriapod as one of its articles of food.

Back of the Pillar of the Constitution is the "Chair
of State"—another handsome mass of stalactites and
stalagmites—that extends from the top of the hill to
the ceiling.   Behind this on the right is the entrance
to "Pluto's Ravine," the roof of which is studded with
representations of sprigs, twining tendrils and branch-
ing corals, all wrought from calcite and "alabaster"
in most exquisite fashion by the hand of nature.
Many are broken, being the remains of those removed
before 1850, when the cave and its contents were esteemed but lightly by the owner, and no care was
taken to prevent its despoliation by visiting vandals.
Beyond this point one can penetrate but a few yards
in the Old Cave, the roof and floor coming close together, and barring further progress.

Much diversity of opinion prevails as to the distance
between the Pillar of the Constitution and the mouth
of the cave.   Stelle, in his work published in 1864,
says it is "just three miles."   Both Collett and Hovey
place it at two miles.   Flint, in 1833, before it was
thought necessary to exaggerate the distance, gives it

as one and a quarter miles, and this is probably not far from correct. The rough character of the passage, the many steep ascents and corresponding declivities, added to the oppressive silence, cause persons unaccustomed to subterranean travels to think the distance much greater than it really is.

THE NEW CAVE.

In 1850 a party from Fredonia, Indiana, observed that a current of air was passing from beneath a large, loosely placed, flat rock at the inner end of Banditti Hall, about 1,000 feet from the entrance of the cave. They succeeded in prying the rock loose and found a narrow descending passage, since known as the "Scuttle" or "Fat Man's Misery." This they entered and passed through, and for the first time white men stood in the "New Cave." The ceiling of the first room entered, which is called "Bat's Lodge," was then black with smoke. Fragments of charred hickory bark strewed the floor, while moccasin tracks, now entirely obliterated, were plentiful. Hundreds of poles of sassafras, papaw, lin, and other soft woods were found both in this room and in that

The Short Route.

portion of Rothrock's Straits nearest the New Cave. None of these poles had been cut with a sharp instrument, but all had been twisted from the parent stem or hacked therefrom with dull stone axes. On the left side of the room was found a sloping bank of earth and sand in which bark, sticks, leaves and bunches of twisted grass were plentiful. Digging into this bank in

November, 1896, numerous pieces of bunch grass, the inside bark of lin and poplar trees and short stems of weeds were found. These were, probably, remnants of a store of fuel resorted to when the torches waned or a relight was needed.

Bat's Lodge is a low room 70 feet long, 21 feet wide and five to six feet high, the walls and roof begrimed with the smoke of ancient fires, and the floor a mixture of dry, dusty earth, with here and there a piece of fallen limestone. From the mouth of the cave to this point is a gradual descension, and barometer measurements showed the floor to be 150 feet lower than the Cave Hotel. Beyond this room the roof so closely approaches the floor that, in 1856, "Counterfeiter's Trench" was dug through the earthy deposit which had silted up the way. Through this trench we easily passed and found ourselves at the foot of "Rugged Mountain," a mass of broken pieces of limestone, thirty feet or more in height, which fills the greater part of a large room. Climbing this mountain we reached the "Rotunda" or upper portion of the room, 52 feet one way by 90 feet the other, with the roof 16 feet above the top of the mass of rock. Around the edges of the room are numerous deposits of fine, white, needle-shaped crystals of epsom salts (magnesium sulphate) encrusting the rocks and sparkling like frostwork in the light of our candles. They seem to exude from a porous matrix of magnesian limestone, and if not disturbed often attain a length of three to five inches. Passing down Rugged Mountain on the farther side we entered "Hanover Chapel," where numerous artificial piles of heavy stones, dedi-

cated to some Greek fraternity or college class, stand
as monuments to the muscular ability of visiting
students in days gone by.

A short distance beyond this point we climbed
again and entered the "Coon's Council Chamber," a
circular room 35 feet in diameter. Here two bands
of blackish flint or jasper about four inches in thickness, first noted in descending Fat Man's Misery, are
very prominent around the walls. A few yards farther on we came to "Delta Island," an uneroded mass
of limestone, 50 feet long by 20 feet wide, on either
side of which one may enter that part of the cave
called the "South Branch," which forms the greater
portion of the Short Route.

Between Banditti Hall and Delta Island a small
Tineid moth, *Blabophanes ferruginella* Hbn., closely
related to the common clothes moth, occurs in numbers. Its presence in such a place is worthy of especial notice since no other instance is on record of a
member of the order *Lepidoptera*, to which belong the
moths and butterflies, being an inhabitant of caves.
This moth was found in May, July and November,

A Cave Inhabiting Moth. close to the floor and always in the
vicinity of the decaying remains of
bats and other refuse matter upon
which its larvæ feed.

It is one-fourth of an inch or more in length, and
its wings expand about two-thirds of an inch. On the
head is a tuft of rust-red hairs. The fore-wings are
grayish-brown, with a violet tinge in fresh specimens,
and a broad buff margin along the inner edge, which, in
repose, forms a conspicuous buff stripe along the back.

The pupal cases are dark gray, densely felted, and one-fourth of an inch or more in length. They were found attached to small projections of the wall, close to the floor, or on the under side of stones which rested loosely on the floor.

The moth seldom flies, but crawls very rapidly or leaps short distances, when disturbed. It was first described in Europe, where it occurs among pelts and furs. Its presence in Wyandotte Cave can only be accounted for by its accidental introduction on the clothing of guides or visitors. As yet it shows no difference in color or structure from open-air types of the same species, but it is not unreasonable to suppose that in years to come there may be perceptible modification in these respects, as has been observed in other cave inhabiting forms.

The introduction of this European moth into a cave like Wyandotte, and its rapid adaptation to the peculiar environment there found, is an excellent proof of the now commonly accepted theory that all cave animals are but the descendants of seeing forms, which in the past, have thus accidentally found their way into caverns.

In the same region in which the moth occurred were numerous specimens of a small dark-brown gnat or fly about one-twelfth of an inch in length. It was found to be new to science, and has since been described as *Limosina tenebrarum* Aldrich. It occurs beneath stones, in the vicinity of the remains of bats which are killed in numbers by the cats which frequent the cave. This insect has the power of leaping several inches when disturbed and seldom uses the wings in endeavoring to escape.

Passing to the right of Delta Island we entered the "Dining Room," forty feet wide, ten feet high and seventy feet in length, the monotony of the limestone walls being relieved by several bands of jet black flint, about three feet apart. One of these bands has the flint in quadrangular blocks, while in the others it is in nodules, many of which are several inches in diameter. Sometimes these nodules resemble in form a geode, and when broken show a crystalline center, the siliceous particles having collected and crystallized about a common nucleus.

Leaving the Dining Room we proceeded through a short pass to the "Drawing Room," whose dimensions are about 25 x 10 x 60 feet, and from this into the "Junction Room." From here three passages diverge, one to the left through "Creeping Avenue," one straight ahead to the right of the "Continent," the latter being a vast mass of uneroded limestone, around which the two branches of the old subterranean river formerly flowed; while the third passage, known as the "Cut Off," turns abruptly to the right and merging into a short, tortuous, descending passage-way, leads out into the main cave between Counterfeiter's Trench and Rugged Mountain.

Taking the passage past the right of the Continent we entered the "Council Chamber," a spacious room, 15x50x100 feet, which, like Hanover Hall, contains many artificial monuments, erected in the past by enthusiastic visitors who knew no better way of proclaiming to the world the fact of their existence. Narrowing again, the main passage continues for perhaps 200 feet, when once more it expands into another of those grand subterranean rooms which characterize Wyandotte Cave. This has been dubbed the "Hall of Representatives," accurate measurements showing it to be 100x160 feet, with the ceiling 35 feet above those masses of fallen rock which in the past filled the space of the broad overhanging dome. Where these large rooms occur, the old river which formed the cave must have flowed over a softer portion of rock and eroded or dissolved a great basin in the bed or floor of the channel, perhaps escaping by an outlet now hidden. In time the roof, no longer self-supporting, came tumbling down and partially filled the basin. From most of the rooms, as from the Hall of Representatives, one must climb twenty or more feet to the mouth of the passage leading onward.

Beyond this hall we descended the "Hill of Science" into a lower portion of the cave, from which a low, wet side passage turns to the right. Here for the first time we encountered mud, and the floor of the "No. 10" passage, as it is called, is for the greater portion of the year covered to a depth of several inches with standing water. We next arrived at the junction room, called "Jordan's Wait," where that noted scientist, Dr. D. S. Jordan, once had several hours for cool reflection, having been left in total darkness by the accidental extinguishing of a candle which he had no means of relighting. This junction room is located at the foot of the Continent, where the passage which turned to the left around that body, meets the one through which we had traveled.

Proceeding onward, we entered the most southern arm of the cave, and, passing through a damp-floored passage, 150 feet long by thirty feet wide, we found ourselves at the foot of a slippery hill on top of which is one of the most handsome formations in the cave— the "Throne and Canopy." The former is composed of a circle of rounded stalagmites cemented together and having the general appearance of a throne of state, while at a distance of six feet above is a curtain of broad, leaf-like stalactites draped in a graceful semi-circle and attached to a projecting mass of crystalline limestone. From a crevice or seam between the massive layers forming the walls the water has, for ages, seeped; then evaporating, has produced these charming natural wonders and given a slippery coat of stalagmite to the surface of the hill below.

In the "Spring of Deception," close by the throne, were noted in July numerous specimens of a small shrimp-like crustacean, *Crangonyx packardii* Smith. It swims very rapidly, jerking itself hither and thither through the water in a zigzag course, and is extremely difficult to capture. In November the water in this spring had disappeared and the bottom was covered with very soft, sticky mud. In this a number of small holes, resembling the burrows or pits of angle-worms, were noted. Each had numerous particles of dry, sand-like grains of mud about the mouth. The pits were probed and cut out with a knife, but no living form could be found. They were probably formed by the small Crangonyx, of which no trace remained. The same crustacean occurs in numbers in Crayfish Spring near the end of the Long Route.

Beyond the Throne is a long stretch of partly explored avenues and side branches, through which

11

visitors are not often taken, there being therein but
one scene of more than passing interest. This is
"Helen's Dome," so named by that Nestor of cave
explorers, the Rev. H. C. Hovey, in honor of his wife.
To reach it one must pass through "General Scott's
Reception Room," 75 by 100 feet in dimensions, and
then by stooping and crawling through a narrow
passage into "Diamond Avenue," where nature
asserts her power to work miracles of beauty from
cheap materials, transforming gypsum and epsom
salts into lustrous crystals which sparkle on the walls
and glisten from the floor. Leaving a branch to the
right, we turned to the left, and passing cautiously
beneath a poised mass of fallen rock, which seemed
ready to fall at the slightest touch, we entered a large
opening midway between roof and floor, and a few feet
farther on found ourselves at the foot of a great circular pit some twenty feet in diameter and extending upward through the solid limestone for eighty feet or more.

This was Helen's Dome, and when the guide kindled
his "red fire," and the light therefrom revealed the
rugged, water-worn carvings of the sides, and the
pendent stalactites, which far above gleamed and
glistened from their inaccessible heights, we with one
accord voted it the wildest and most romantic bit of
scenery which the cave possessed.

Retracing our steps to Jordan's Wait, we took the
right branch around the Continent. This led us on
through a low passage known as "Purgatory," 140
feet in length, its floor of yellow ochre, with here and
there a handsome crystal of selenite; its roof of white
limestone, with many fantastic grooves and carvings

wrought in days of yore by the slow but powerful energy of flowing water.

Emerging from Purgatory we assumed once more a standing posture, and found ourselves in "Calliope's Bower," where many stalactites grace the walls and ceiling. From thence we passed into "Whispering Gallery," where the floor resounded to our tread and the low tones of our voices were echoed back and forth from the arched sides in a manner similar to that noted at the bottom of a deep and empty cistern. Then came the "Palace of the Genii," where these gods of fable dwell beneath a roof spangled with glittering crystals of calcite and gypsum. The "Pillared Palace" follows, and therein is found a wealth and profusion of cave formations such as no words of man can properly picture. Pillars, stalagmites and stalactites abound of every conceivable form which the fancy can suggest. Many of the stalactites are no larger in diameter than a lead pencil and are curved and twisted in a unique and grotesque manner seen elsewhere in no Indiana cave. This bent and twisted condition is doubtless due to the varying currents of air which pass through portions of the cave and force the tiny drops of water on the end of the stalactite first to one side and then to the other of the tip. The air of Wyandotte flows outward, or toward the mouth, in summer, and inward, or toward the depths of the cave, in winter. This difference in direction of flow can but have its influence on the formation of such slender structures as those above mentioned.

Emerging from the Pillared Palace by an ample doorway, flanked by handsome pillars of calcite, we found on our left a room where strata of jasper nodules abound in the walls and where numerous chips and splinters of jasper are abundant on the floor. Rev. H. C. Hovey first called attention to the fact that the supposed "bear wallows" of this room are depressions where, in the treacherous light of bark torches, ancient workmen had reclined while they worked down to partial finish the desired blocks of jasper. Numerous fragments of charcoal and large heaps of chips of jasper were about each depression, but, though careful search was made, no partially finished article of jasper was found. The fragments were mostly oblong, with the faces parallel, their dimensions being, on an average, about $4 \times 2 \times \frac{1}{2}$ inches. Several quartzite bowlders have been found in the room, where they were doubtless carried to be used as implements in splitting the blocks of jasper or in loosening them from the walls.

This ancient quarry-room is succeeded by "Creeping Avenue," where the roof, for a distance of 172 feet comes down to within two and one-half feet of the floor, and progress is possible only upon the hands and knees. According to the guide, the dryness of this portion of the cave is slowly increasing and, as a consequence, epsom salts (magnesium sulphate) is becoming more abundant. Where the cave is damp with dripping water, stalactites and other forms of calcium carbonate are abundant; where the dripping has ceased but the walls still give off more or less dampness, calcium sulphate or gypsum is the prevailing formation, and where perfectly dry the epsom salts alone are being produced.

The tiresome crawl through Creeping Avenue finished, we stood erect once more in the Junction Room at the head of the Continent and the exploration of the Short Route was at an end. The length of the portions passed through was estimated to be about as follows:

Fat Man's Misery to Delta Island .... 1,200 feet.
Delta Island via Creeping Avenue to Hovey's Point ..... 2,400 feet.
Jordan's Wait via House of Representatives to the end of
    the Cut Off. .... 2,000 feet.

    Total .... 5,600 feet or 1.06 miles.

In going through what is known as the "Long Route" in Wyandotte, we passed from the mouth of the cave to Delta Island over the same way as described above under the "Short Route." At Delta Island we turned to the left and traversed the "Sandy Plain," a passage about 350 feet long, twenty-five feet wide and six to ten feet high; the floor of which is covered in places to a depth of several feet with sand deposited

The Long Route. by the ancient cave river. At the end of the Plain we found ourselves at the foot of the "Hill of Difficulty," which is but a mass of fallen rock, forming, as it were, a foot-hill to the grander "Monument Mountain" which lies beyond. On the left, in climbing this hill, the guide pointed out the exit of Rothrock's Straits, that narrow and deeper passage connecting the Old and New caves.

Reaching the top of the Hill of Difficulty, we were within the confines of the largest under-ground room yet known to man—"Rothrock's Grand Cathedral."

Before us in the dim candle light was a towering mass of fallen rock, thrown together in most glorious confusion and piercing the gloom above us for 135 feet. Following the guide and clambering from rock to rock, we made the ascent by easy zigzags and reached a point near the summit with but little fatigue. The crest of Monument Mountain, like that of Capitol Hill in the Senate Chamber of the Old Cave, is covered to a depth of several feet with an encrustation of stalagmitic material. This is slowly increasing in thickness by the accretion of solid particles of limestone left by the evaporation of the water which is constantly trickling in a small stream from the roof above. The uppermost ten or twelve feet of the mountain is very smooth and slippery, and one has much difficulty to keep his footing while climbing to the very pinnacle, from which projects a brownish-yellow stalagmite 6.5 feet in height and 3.7 feet in circumference. Below this a short distance, and on the opposite side of the mountain from the entrance, is another stalagmite, 6.8 feet in height by 5.2 feet in circumference, while but a short distance away is a third and shorter one. The last two are composed of spotless white, almost translucent limestone, and are known as "Lot's Wife and Daughter."

Crawling over the damp surface at the foot of these stalagmites, as well as on their sides, were numerous specimens of cave myriapods and harvestmen. A few examples of a small, semi-blind pseudo-scorpion, or chelifer, *Chtonius packardii* Hagen, were also obtained from the surface of the damp rocks at this place. It moves slowly along with its front legs or

XIII

1. A Pillar in Pillared Palace, Wyandotte Cave.
2. Rothrock's Cathedral, Wyandotte Cave.

chelæ held in the air and, being less than one-tenth of an inch in length, is very likely to be overlooked unless especial search is made for it. It has been taken in Mammoth and other Kentucky caves, and varies much in regard to the development of the eyes; some, living without the caves, having two eyes with the cornea as usual; others, within the caves, having no cornea, but retaining the silvery dot indicating the retina, and still others being totally blind.

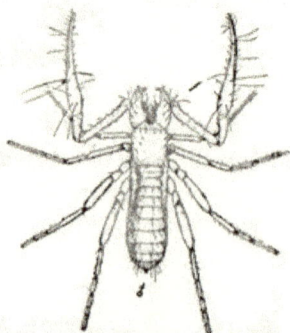

Fig. 35—Cave Pseudo-scorpion. (After Hubbard). (Enlarged 7½ times.)

Forty or more feet above the crest of Monument Mountain expands "Wallace's Grand Dome." The center piece of this is, "a smooth, elliptical slab of oölitic marble, 60 feet long by 30 wide, finely contrasting with the darker limestone, from which it is divided by a deep rim, fringed with long stalactites, curling like leaves of the acanthus."

Leaving three candles burning a few feet below the summit, we descended the opposite side, and, extinguishing our lights, as soon as the eye accustomed itself to the surroundings, beheld a scene as grand as human mind can fancy—"an indescribable vision, as if an opening had been made into the realms of supernal splendor." The scene is known as the "Cathedral by Moonlight," the faint candle light reflected from the white, oval dome appearing like a halo of moonlight over the dark crest of the mountain, while the three stalagmites stood like spectral visions surmounting the dark and rugged ledges which rose between us and the source of the faint light above.

Relighting our candles, we found a few feet farther on, the "Sulphur Spring," the trickling waters being caught in a shallow cavity of a round stalagmite. This is one of the dining places of visitors, and from the near-by moldy remains of food were taken numerous specimens of a small black fly, *Phora nigriceps* Loew, as well as several mites, and the common cave "springtail."

To the left of Sulphur Spring, in the smooth and slippery stalagmite, is an opening thirty inches wide by fifteen inches high. This is the famous "Auger Hole," which, when discovered and enlarged in 1850, admitted the explorers to an extensive area of unknown passages and rooms—yet none so grand as those already noted.

Through this opening we made our way, some head first, others the reverse, all finally landing safely about ten feet below in a damp room known as "Lilliputian Hall," along which, by stooping, we found our way into "Spade's Grotto," once evidently connected with Rothrock's Cathedral by a passage now hidden by fallen rock. From thence, in divers manners, we descended "Slippery Hill" and found ourselves in the "Hall of Ruins," a passage 150 by 30 feet, with an average height of perhaps eighteen feet. This leads into the "White Cloud Room," probably 350 feet in length, where the roof and walls are encrusted with an efflorescence of gypsum, resembling, after a fashion "billows of fleecy clouds." Beyond this room we passed through the "Journal Office," near the farther end of which is the "Bishop's Rostrum," a high platform of rock, 8x10 feet in dimensions, from which portions of many a sophomoric oration, as well as several divine dissertations, have, in the past, been delivered.

"Calypso's Island" is a large mass of uneroded limestone, on both sides of which the old cave stream has forced a passage. The floor of the passage to the left of the "Island" resounded our footsteps in a peculiar echoing fashion, suggesting the presence of a lower passage beneath our feet. The two wings of the main passage converge at the farther end of Calypso's Island and expand into the "Cærulean Vault," a room 40 feet wide by 20 feet high. This narrows into "Rugged Pass," from the side of which a narrow cleft in the rock leads by an ascending, very low and tortuous passage, known as "Worm Alley," into "Milroy's Temple."

This is a room 100 x 150 feet in dimensions, around the upper edge of which are found some of the most handsome formations in the cave. One of them is a row of musical stalactites, broad and thin, on which a melody can be played by a skillful hand. There are also creamy stalactites, vermicular tubes strangely intertwined, convoluted roots, mural gardens and galleries, gay and grotesque. A deep pit, the bottom of which is sixty feet or more below the entrance, is found in one side of the room, and the sound of a stream of water falling from a cleft in the ceiling and splashing on the rocks at the bottom of the pit was a pleasing break to the monotonous silence of the vast rooms through which we had come.

Once more bowing our heads to the inevitable, we crawled, squirmed, rolled and pulled ourselves through Worm Alley back into the main cave. Following our guide, we passed on through "Josephine's Arcade," where a silhouette of the "Cave Queen," formed by the falling away of the white gypsum from the darker limestone, greets us from the wall. "Indiana University Chapel" and the "Ball Room" succeeded and brought us to the "Junction Room" of the Long Route. Here the cave forks, one branch leading to the south-west and the other continuing northward to "Crawfish Spring" and "Wabash Avenue." Taking the latter, we found it to be made up of a succession of halls, galleries and avenues, each with its own fanciful name and pleasing peculiarities, yet no place worthy of more than passing notice when taken in contrast with the grand scenes already described.

In several places between the Junction Room and Crawfish Spring the first explorers of the Long Route found tracks of a small party of Indians who had wandered to and fro in that region. They had evidently entered by some opening as yet unknown; since the Auger Hole, now the only means of entrance, was, when first discovered, entirely too small for the passage of a man. It is better, in my opinion, to consider that their means of entrance and exit has since been covered by fallen rock or, like that through Fat Man's Misery, was hidden purposely by those ancient explorers, than to take the ground, as has been done, that the tracks were made 1,800 or 2,000 years ago, before the opening of the Auger Hole was so nearly closed as to prevent the passage of a man. These

moccasin tracks were seen and noted by many of the
early explorers, and low stone walls were put around
them for protection, but the tracks have since been
almost entirely obliterated by persons who, unmindful
of the warnings of the guides, stepped over and upon
them.

Crawfish Spring is formed from a small stream
which flows through a cleft in the rock, and from it a
trickling rill meanders on beneath the edges of the
jutting walls to be soon lost to view beneath the roof
which a few rods farther on comes down to meet the
floor. Above the spring is the passage known as
Wabash Avenue, which extends for several hundred
yards in a north-westerly direction where it forks into
a number of low and muddy branches.

Within the waters of the rill were several specimens
of the blind crawfish and numerous examples of two
other smaller crustaceans, already mentioned. About
the margins of the spring and stream and on the mud
flats lying beyond, were secured a number of the true
cave beetles, *Anophthalmus tenuis* Horn. Single specimens of this insect had previously
been taken in several of the caves
visited, and in Wyandotte it had been
found about the Throne and on top of Monument
Mountain. It is found only in remote parts of
the caves in which it occurs, and is always crawling rapidly over mud, sand or rocks in damp localities. It is a small, light-brown species, with no ves-
tige of eyes, and appears wholly unaffected by the
light of a candle when the latter is held within a few
inches of it.

A Blind
Cave Beetle.

Like other Carabids, these small blind beetles are supposed to be carnivorous. In Wyandotte specimens of mites, spiders, spring-tails and harvestmen were taken in the same locality as the beetles, and probably furnish the latter a scanty supply of food.

Fig. 36—Blind Beetle.
(After Packard.)
(Enlarged 3½ times.)

Retracing our steps to the Junction Room, we took the south-west passage, the first room entered being the "Frost King's Palace," eight feet high and twenty wide, where every object, great and small, is encrusted with sparkling crystals of gypsum. To one side is the "Bridal Chamber," and therein are found some of the finest of the gypsum rosettes for which the cave is noted. Several of these are four and a half inches in diameter, the slender crystals forming them having protruded from the pores in the magnesian limestone, and then, uniting into fibrous masses, have curved inward to form the *oulopholites*, or curl-leaved stones, each of which bears a close resemblance to a true rosette.

The "Ice House" is a rough-floored room where dripping water from the roof has covered the surface of the rocks with a film or coating of the thinnest and most translucent of calcite, resembling ice. Leaving the opening to the "Unexplored Regions" on our left, we descended from the Ice Room into "Morton's Marble Hall," 1,100 feet in length, the sides and walls of which "are completely dressed in snowy whiteness, equaling the brightest marble halls of dreamland, song or story." Occasional nodules of jet-like flint are seen exposed along the walls and ceiling, and here and there are examples of the gypsum rosettes already mentioned. Beyond the Marble Hall is "Queen Mab's Marble Garden" and the "Fairy Palace," both of which have their walls covered with a gypsum efflorescence which has assumed the shapes of flowers, leaves, sprigs and fanciful forms of many kinds. From the end of Fairy Palace, 1,750 feet from the Ice House, diverge several low passages which visitors seldom enter, and from here we started on our return to the entrance of the cave.

The distance from Delta Island to Crawfish Spring, including Milroy's Temple, was estimated to be about one and four-tenths miles, and from the Junction Room to the end of Fairy Palace, about one-half mile. The total length of Wyandotte Cave as traversed by the visitor who takes all three routes is, therefore, about as follows:

Old Cave	1.25 miles.
Short Route from Fat Man's Misery onward	1.06 miles.
Long Route from Delta Island onward	1.90 miles.
Total	4.21 miles.

These distances are for one way only, and if the routes are passed over on different days, the distance from the entrance to where each begins must be added to that above given.

With two of the guides I passed through Rothrock's Straits in November, 1896. Dropping ourselves through the narrow cleft in Odd Fellows' Hall, we crawled down an angling passage over a mass of rough rocks and into a low room almost filled with

fallen rocks. From this we crawled still farther
down, climbing over great blocks of limestone and
making our way beneath others partly loosened from
the roof, until finally we reached the
very bottom, probably seventy-five feet
below our starting point. Here we
found another low room, with an earthen floor, which
had great cracks running through it in every direction, but with no signs that water had been present
for centuries. From this a very low passage makes
its way to near the Coon's Council Chamber, but
there is no exit into that room. Retracing our way
we took another route, and after much creeping,
wriggling our way through dust, bumping our heads
on the low ceiling, and with nothing in the way of interest to repay us for our trouble, we finally emerged
on the side of the Hill of Difficulty, and knew by
personal experience that the Old Cave and the New
are connected, and that the passage-way between
them is a very rocky road to travel.

Rothrock's Straits.

The entrance to the so-called "Unexplored Regions"
of Wyandotte opens from the Ice House, beyond the
Junction Room of the Long Route. These regions
have been partly explored by guides,
but visitors seldom pass within their
portals. Washington Rothrock, the
oldest and best known guide of the cave, has been
through them several times as far as Rothrock's
Island. The formations therein are said to be won-
drously beautiful and more numerous than in many
of the more thoroughly explored regions of the cave.
A large series of specimens were obtained from some

**The Unex-
plored Regions.**

of these passages in 1893 for the World's Fair. A number of the larger passages of the Unexplored Regions have not been penetrated as far as man can go, and some future explorer may, perhaps, find formations more beautiful and scenes more grand than those occurring in the better known portions of the cave.

LITTLE WYANDOTTE CAVE.

The entrance to this cave is situated at the bottom of a sink-hole distant about 300 yards from the front of Wyandotte Cave Hotel. The floor of the cave is about 20 feet below the bottom of the sink, and descent is made by a ladder placed in a well-shaped opening about three feet in diameter.

In the crevices on the sides of this opening were several cave salamanders, and also a number of the large hump-backed cave crickets, *Ceuthophilus stygius* (Scudd.). This insect reaches a length of one and a fourth inches, and has antennæ, or feelers, more than four inches long. It is not a cricket, but belongs to the same family as the katydid,

A Cave Cricket.

though in general appearance it differs widely from that common insect. It was found in the entrance of Wyandotte Cave, and in several other of the smaller caverns of the vicinity, but in no instance farther back than 250 feet from the mouth. The adults seem to be more or less gregarious, and, in one instance, more than 20 were found in a small cranny in the wall. They were grouped in a circle, in a space about six inches square, with their long antennæ pointing toward the center of the circle, and appeared to be holding a conference or cricket convention.

They were never seen on the floors of the caves unless they leaped there when disturbed, but were found resting on the sides of small projections and in cavities of the walls or roof. If a lighted candle were held close to them they paid no attention to it, but were very sensitive to its heat and to touch. When disturbed they leap with agility, sometimes to a distance of six feet, but with a little care can usually be readily picked up with the fingers before they become frightened. Being wingless they make no noise, and like most other silent creatures, are supposed to be deaf, as no trace of an ear-drum is visible.

At the bottom of the opening into Little Wyandotte one finds himself in an entry which leads both to the right and the left. The right hand passage can be followed only about seventy-five feet, when it becomes too small for farther progress. It contains no features of interest except a few stalactites.

The left hand passage was found by actual measurement to be 340 feet in length. Passing "Pompey's

MAP OF LITTLE WYANDOTTE CAVE.
Crawford County, Ind.

100 Fr

Entrance.

1 and 2, Pits.   3, Gallery.

Fig. 37.

Pillar," a large stalagmite, the first room entered was "Cleopatra's Palace," where there are hundreds of fine stalactites, which show grandly in the glare of the magnesium light.

Beyond this room two pits, said to be 60 feet in depth, shut off the farther side of the cave. A narrow partition of slippery stone separates the two and serves as a bridge to cross the chasm. On leaving this natural bridge, we made our way along the side of a steep ledge that skirts the left hand pit, and then passed around a gigantic fallen stalactite, which has been kept from rolling into the pit only by a friendly stalagmite against which it rests. Climbing a steep slope in which notches have been cut to serve as footholds, we entered a gallery, one side of the expanding mouth of which serves as a balcony above and partially around the deepest pit. On and above this balcony is a collection of cave formations of exceeding beauty and grandeur. A stately, fluted pillar, with its base expanding in broad-leaved masses of dripstone, thus forming a heavy folded curtain along the edge of the pit, is the giant of the group; while most unique of all is the "Corinthian Column," ten feet high and less than three inches in diameter—a slender shaft of translucent, snow-white satin-spar reaching from floor to ceiling. A number of fragile tubular stems were clustered about the head of this pillar, each with a terminal drop of water, which glistened like a well cut diamond in the light of our candles. Entering the gallery we wandered on, "beneath a ceiling fretted with glistening pendants, amid pillars and pilasters, flying buttresses and interlacing arches, with here a

12

cascade in mid-air transmuted into stone, and there a
sculptured cell amid clustered columns." The cave
finally ends in "Peri's Prison," where a narrow side
gallery is separated from the main passage by a row
of slender pillars, each but a few inches from its
neighbor. All in all, Little Wyandotte is well worthy
of visitation, and he who wishes to see the beautiful
and at the same time experience a sense of the perils
attending cave exploration, should enter its bounds,
cross the narrow bridge between the yawning chasms,
and climb the slippery hill to the lovely gallery beyond.

* * *

Other caves there are in southern Indiana which we
would gladly have explored and described had our
time permitted. No two in the State are alike. Each
is noted for some peculiar formation, room or passage
which it possesses. In each and all can one see the
results of the action of water—that greatest of nature's
solvents and abraders, soft to the touch, gentle to look
upon, its work of a day, a year, a century upon the
solid limestone not appreciable to the eye—yet by
slow, unceasing action through the eons which have
elapsed since that work began, it has carved every
room and passage, constructed every pillar and stalagmite existing beneath the surface of southern
Indiana.

A DAY IN A TAMARACK SWAMP.

A half-day rather, for the other half was occupied in reaching the swamp, and half of the night in returning therefrom, so that part of the story must be of some of nature's objects noted on the way and the thoughts that they engendered.

The 6:20 train on the "Logan" road was almost ready to pull out on a recent Saturday, when, after a brisk walk of half a mile through the cool enlivening air of early morn, I reached the Union Station. The platform was thronged with prospective passengers, bound for the four corners of the globe and starting with the coming of the sun. All was bustle and noise but we were soon off to the quiet prairie region north-east of Terre Haute.

No frost as yet had seared the vegetation and the late planted corn was ripening rapidly; its partially green, partially yellow leaves glistening with the dew of the night. The many wild plants growing along the railway are to me, when traveling, objects of exceeding interest, but on this day but few were at first seen, the ruthless hand of the section boss having caused their early downfall. Occasionally a sunflower or stalk of golden-rod, which in some unknown way had escaped destruction, waved its flowers in defiance as we were whirled past; or the bright blue of the wild morning-glory and brighter blue of the lobelia

(179)

seemed trying to excel the hue of the sky above them.

Just over the boundary fence, however, beyond the jurisdiction of the devastating wayside mower, grew in many places wild asters in profusion. Our native asters are distinctively flowers of autumn. They do not begin blooming until mid-September, and, as late as December 1st, can often be found in some protected nook, the last wild flowers of the dying year. The wild asters vary in color from a pure white to a deep blue. One or two possess a pinkish tinge but none are red or yellow. Our sunflowers and golden-rods furnish sufficient of the latter color; while the scarlet leaves of the maple, black-gum and dogwood, together with many of our wild fruits, paint amply red the autumn landscape.

The asters, sunflowers and golden-rods comprise nearly one-half of the Indiana members of the great *Compositæ* family—in number of species the giant family of the flowering plants. In the arts the use of these plants is unknown. But the lover of nature, whose eye is ever on the search for the pleasing and the beautiful, blesses their existence; for, their hues absent, our autumn scenery would lose much of the charm due to its variety of color.

Beyond the prairie came the uplands of Parke and Montgomery counties where the Kentucky blue-grass and the sugar-maple, two of the most attractive and valuable wild plants known to man, reach the acme of their perfection. The limestone soil of this "blue-grass region of Indiana" furnishes exactly the food needed to make the sugar-maple, *Acer saccharum*

Marsh, a perfect tree. Great groves of them, each tree so isolated from its fellows as not to shade too densely the grass beneath; each with its branches stopping short at a uniform height, ten to twelve feet from the ground, form vast forest sheds which all summer long furnish plentiful shade to those herds of fat cattle which are the pride and wealth of the owners of the land.

Farther north the "Logan" runs through a flat, upland country where the bitter-nut or swamp hickory, the beech with its smooth lichen-covered bole, and immense numbers of gigantic bur-oaks, abound. At intervals clumps of that handsome shrub, the black alder or winter-berry were seen, its bright-red fruit giving an exceeding vividness to the dense green of the surrounding forest, which was as yet untouched by that prince of painters, Jack Frost.

Then the prairie with its characteristic flora once more appeared; and finally, the tamarack swamps about Kewanna, Fulton County, came into view and my journey by rail was at an end.

The tamarack or black larch, *Larix americana* Michx., is a tree, fifty to one hundred feet high, with a straight trunk and slender horizontal branches. It belongs to the great Pine family or *Coniferæ*, so called because the seeds of its members are borne without

The Tamarack Tree. other covering than the large flat scales which overlap one another to form that familiar object called a cone. The large majority of the members of the family, as the pines, cedars, etc., are evergreens; but the leaves of the tamarack, which are thread-like, one or two inches in length, and arranged along the branches in small bunches, wither and fall each autumn. The cones of the tamarack are about one-half inch long, ovoid or egg-shaped, and purple or brownish when they ripen. The wood is hard, strong, and very durable, and is used for ship building, fence posts, telegraph poles and railway ties. The slender roots are composed of tough fibres of great length. The Canadians and Indians were accustomed to use these fibres for sewing their bark canoes, hence Hiawatha is said to have made the following request :

"Give me of your roots, O Tamarack!
  Of your fibrous roots, O Larch-Tree!
My canoe to bind together,
  So to bind the ends together,
That the water may not enter,
  That the river may not wet me."

The tamarack is a lover of cooler climes than is furnished by our latitude, and hence flourishes in greatest abundance in the far north, its southern range in this State being on a line drawn east and west through Ft. Wayne and Kewanna. Like its cousin, the bald cypress, its chosen home is low, swampy land where it often thickly covers large areas and furnishes that dense shade so characteristic of any member of the pine family where growing closely together in great numbers.

After entering a few yards within the swamp a sense of solitude and loneliness, such as I have never felt, even in the most dense of our ordinary forests, began to oppress me—a sense which increased with every onward step and did not wholly disappear until

I had come forth again into the full light of the sun
and had left far behind me the swamp with all its
characteristic surroundings. Much of this feeling
was doubtless due to a lack of those animal sounds
usually present in a forest. This lack may not be so
noticeable at other seasons of the year, but on this
September day it was especially striking. I listened
in vain for the chirp of bird or the hum of insect.
The silence was broken only at long intervals by
the trill of a striped tree frog, or the low soughing of
the wind in a weird and mournful cadence through
the thick branches of the trees about me. The lower
limbs of the tamaracks, which hung downwards as
if weeping, were, for the most part, lifeless and covered with gray lichens. Many of these lichens grow
in long and slender tufts like the "Spanish moss" of
the Southern States; this sombre drapery but adding
a deeper tinge of desolation to the scene.

But, however lonely such a swamp may appear to
one who traverses it, to the botanist it is ever a place
of interest on account of the many rare plants which
lurk within its bounds. No grass or sedge can exist
in the profound shade of the trees, but three or four
kinds of sphagnum mosses grow everywhere in deep,
dense masses which gave way like windrows of new
mown hay beneath my tread. When dry these beds
of moss furnish delightful cushions, on which when
tired, one can throw himself down and rest at ease. I
dug deep into one of them but could find nothing but
layers of moss, the older stems slowly decaying, the
younger finding a foothold and sustenance among the
ruins of the old.

Scattered through and over these mossy beds were many trailing stems of a slender shrub, bearing oblong, evergreen leaves about one-third of an inch in length. Very handsome these shrubs were, and interesting too, for on them grows that delicious and familiar fruit, the American Cranberry, *Vaccinium macrocarpon* Ait. The berries had been carefully gleaned from the swamp, but here and there the bright cheek of one glistened from its bed of green, furnishing a natural contrast of color which would entrance an artist's eye.

At intervals of a few feet among the thickest of the tamaracks were clumps of that curious carnivorous growth, the side-saddle flower or pitcher-plant, *Sarracenia purpurea* L. The margin of its thick root leaves

Fig. 38—Common Pitcher Plant. (After Bessey.)
(One leaf cut across to show the cavity. One-third natural size.)

are united in such a way as to form hollow tubes or "pitchers" with a rounded lid or lip at the top. On the inner surface of this lip are numerous stiff hairs pointing downwards. A watery fluid is secreted by the leaves, and collecting in the pitchers, attracts many insects. These are soon drowned, being unable to escape from the fatal pitchers on account of the deflexed hairs, and their bodies, decomposing, yield a plentiful supply of nitrogen for the crafty plant.

In May and June the strikingly handsome flower of the pitcher plant, deep purple in color and two inches or more in diameter, may be seen by the visitor to these swamps. It is borne on the summit of a leafless flower stalk or scape which springs from the midst of the clump of pitcher-shaped leaves.

In one part of the swamp grew in abundance wild huckleberries, own cousin to the cranberries, and in their season, in high favor for pies and cobblers; yet for a sauce, a failure, on account of a lack of sufficient juice and acidity. Orchids, too, were there, noted for their curiously formed flowers, which, like those of the milk-weed, can only be fertilized by the aid of insects. Though past their blooming season, three or four species were noted, among them the yellow lady's slipper, once common throughout the State but now almost extinct on account of the ruthless, devastating hand of man.

Two slender shrubs, each four to six feet high, grew in such open places as occurred. One, the choke-berry, *Pyrus arbutifolia* L., is a close relation to the apple, but bears its fruit in clusters like those of the black haw of our woods. The fruit is in shape like a miniature apple, dark red or blackish in hue, and edible, but with a puckering taste like that of a green per-

simmon. The other is the dwarf birch, *Betula pumila*
L., a northern plant which, like the tamarack, reaches
in this swamp its southern limit in our State.

One other plant which I fain would omit from the
list of those noted must now be mentioned. All those
preceding are either useful or harmless in their ways,
but this one is the most poisonous species known to
the flora of Indiana. It is the poison sumach or swamp
elder, own brother to our common ivy but much more

**The Poison
Sumach.**

venomous as its scientific name, *Rhus
venenata* DC., denotes. It is a shrub
growing to the height of twelve or
more feet. Its large compound leaves are often
two feet in length and composed of nine to thirteen
slender leaflets; while from their axils the white,
grape-like, fruit hangs in loose bunches. It grows
only in the northern swamps and its juice, or even the
exhalation from its leaves, causes small white blisters
to appear anywhere on the surface of the exposed
skin. In this swamp it was abundant, growing in
every open space, and although I tried to avoid it as
much as possible yet the blistered, itching skin which
I endured for days after my return proved to me too
well its poisoning powers.

The scarcity of animal life within the swamp has
already been mentioned. It is said to be a famous
place for owls, and is in every way well fitted for
those ominous birds of prey which delight in all that
is dark and dismal. In the words of Thoreau:
"Their '*hoo-hoo-hoo, hooer-hoo*', is a sound admirably
suited to swamps and twilight woods which no day
illustrates, suggesting a vast undeveloped nature which

men have not recognized. They represent the stark twilight and unsatisfied thoughts which all have." I saw no owls, but the harsh caws of a distant flock of crows which at times were wafted to my ear told where one was treed and pestered; for the crow hates an owl as badly as a terrier hates that other "bird of night," the prowling cat.

The mottled grasshopper, *Melanoplus punctulatus* (Uhler), frequents the depths of the swamp in small numbers, resting either upon the trunks of the tamarack trees or the clumps of sphagnum mosses at their base. It is not active in its movements, usually, after one or two short leaps, squatting close to the earth and, seemingly, depending upon the close similarity of its hues to the grayish lichens about it to avoid detection.

In one of the drier parts of the swamp a prairie rattle-snake gave its shrill warning almost beneath my feet, and mosquitoes of the large striped variety, regular "gallinippers," as the boys call them, occasionally had a tendency to taste my flesh; but these were only minor drawbacks which every wooer of nature must at times endure, if he would see her odd corners as well as her more commonplace ones.

* * *

I wrote the gist of the above while sitting deeply in a bed of sphagnum moss, so deep indeed that I could scarcely move my arm while writing. And O, how tired I was! For 140 miles of railway travel and a five hours' tramp over bog and tussock will tire the average human.

How I longed to lean backwards and sleep! But the rays of the sun, slanting and struggling beneath the lower branches of the tamaracks, warned me that my train would soon be due, and that if I wished to sleep at home, safe, except in dreams, from the attack of rattle-snake and mosquito, I must be up and away for the mile's trudge to the station. And so my day in a tamarack swamp became to me a thing of the past.

MID-AUTUMN ALONG THE OLD CANAL.

Old Mother Earth once more had made the circuit of the sun and October, the fairest month of all the year, had come again and brought with it one of its perfect days. The chilling winds and hoar frosts of the week before had warned the Red Man in the far west of the approach of winter and, for the first time this season, he had kindled his signal fires, and from them the smoke,

> "Soft and illusive as a fairy's dream,
> Lapped all the landscape in a silvery fold."

On such a day the gypsy in my blood—that desire to roam and wander which I inherit from the barbarians of old—asserts full strong its presence. The city with its crowds and turmoil, its noisome smells and impure atmosphere, becomes for a time unbearable. Only a tramp through field and forest and a communion with some of the many spirits of the woods will serve to curb this gypsy element and give me peace of mind once more. And so on this, the sixteenth day of the month, and hence the very middle of the autumn, I started northward, I knew not whither; I cared not whither; but the old canal proved a cynosure and the spirits with which I communed are in part noted below.

Birds, birds, always to be seen as soon as the city limits are reached, always interesting always full of life and sprightliness. A bluebird first. Then a flock of juncoes or snowbirds, fresh from their summer home in the far north, perhaps tarrying for food before they go farther south, perhaps come to stay with us, to cheer us up with their chirp and twitter when King Boreas with his attendant train of ice and snow, of bare trees and almost voiceless nature, will rule over us. Then the familiar song sparrow dodging from fence crack to brush pile and back again, his streaked breast ever a sign of his identity. Then a company of grass finches showing the pure white of their tail feathers only as they fly; preparing for their southward journey by thus flocking together from their various nesting places. Then a sound—"*ha-ha— ha-ha*"—and a trio of crows, uttering their weird laugh, went sailing southward, seeking, no doubt, their morning meal on the commons near the city. Next, a bevy of meadow larks were flushed, and flying across the road they glided, as it were, down an inclined plane until they reached the ground, much as a flying-squirrel travels from tree top to the earth below.

By this time I had reached the canal and within the confines of its banks, where the killing effects of the recent frosts were not so visible, birds were plentiful. A pair of white-bellied nuthatches ran industriously to and fro on the branches of an elm beneath which I rested. They peered into every cranny and looked beneath every piece of loose bark in their careful search for the luckless insects which were destined to serve them for dinner. At short intervals they talked to one another in two brief words, "*kah— kah;*" when frightened, repeating them very rapidly, "*kah-kah—kah-kah.*" At times, also, a kind of low. chuckle or "*pit*" like sound was heard as though the bird had suddenly thought of something pleasing and was laughing to itself.

The "*pe-a-body, pe-a-body, a-body, a-body,*" long drawn out, of the white throated sparrow, mingled with the much louder and harsher "*che-wink*" of the

Fig. 39—Meadow Lark. (After Beal.)

marsh robin or tow-hee came from the many brush piles along the sides of the canal; while the rapid "rat-tat-tap" of a downy woodpecker upon a dead snag furnished the bass for this medley of bird sounds.

Do the members of the different families of birds understand and converse with one another? It would seem so, for whenever a cry of distress comes from a wounded or frightened bird, species of widely different families flock about and, with varied notes of alarm evince an interest in and a sympathy for the woes of their unfortunate companion of the woods.

From mid-August to October

> The ceaseless hum of insect life goes ever on,
> No pause for night or morn or noon-day sun.

After the first frosts, however, these insect sounds grow gradually less until November, when they almost wholly disappear. On this day the chirp of a wayside cricket, the crackling note of a clouded grasshopper, made by the male while on the wing, the drawling call of a harvest-fly which had long out-lived its day, and the feeble shrill of two or three small species of katydids were the only insect notes which were heard.

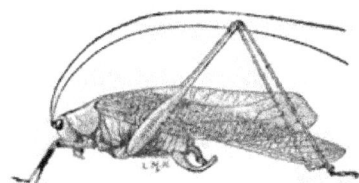

Fig. 40—A small Katydid. (After Lugger.)
Scudderia furcata Brunner.

Even they were only occasional weak wails —woe begotten sounds of frost-bitten individuals—and not the loud shrill notes of the same species of a fortnight before, when love demanded tribute in the form of ceaseless song and all went merry as a marriage bell.

Where a barbed-wire fence stretched across the bed of the canal a novel sight came into view. Hundreds, yes thousands, of strands of spider webs were floating from the sides of the posts and wires. The wind was from the south and they were blown northward, the free ends floating horizontally and parallel in the air.

Spiders As Balloonists.

A close examination revealed the presence of a great many spiders of minute size on the surface of the posts and wires. Their abdomens were slightly raised and from the spinnerets of each were issuing threads of web material which soon hardened when exposed to air. These, in time, formed a strand sufficiently long to support the weight of the little spinners. Then, if the strand did not become entangled with some others, the spiderkin, letting go its hold upon the fence, took a ride to the northward with the wind as a motor. Wonderful balloonists they—whose ancestors performed like feats of aërial navigation ages before the days of the first human aëronaut!

Numerous wild fruits and nuts, both edible and unedible, are in autumn found along the banks of the canal. The former satisfy the sense of taste and are usually dull in color but interesting in structure and habit. Among them are two species of wild grapes, one, the frost or chicken grape, *Vitis cordifolia* Michx., with small black and shining berries which are very sour; the other, the summer or fox grape, *V. æstiralis* Michx., having the berries larger, with a sweet and pleasant flavor and with their black skins covered with a whitish bloom. The chief charm of the wild grape lies, however, in the spreading, straggling habits of the vine which covers many thorn and other homely bushes, and forms in the angles of the old Virginia rail fences those dense, leafy coverts which in summer delight the brown thrush and make glad the heart of the handsome chewink.

The hazelnut, *Corylus americana* Walt., also flourishes in this sandy soil and on this day many clumps

13

of them were noted. Their oval leaves painted a dainty brown by the frost, were withering and dropping, thus disclosing more plainly the bunches of brown nuts each clad in its protective armor of involucre. An interesting fact about the hazelnut is that its catkins, earnest of next season's flowers, are formed in late summer and pass the winter in patient waiting, ready to take advantage of the first warm days of spring to open their cups of pollen and fertilize the flowers for the future crop of nuts.

Our unedible wild fruits usually cater to the sense of sight, being bright in color or peculiar in structure. Among them all, that of the wahoo or burning bush seems to me most beautiful. Hanging on slender pedicels, four or more in a cluster from the same peduncle, its deep scarlet color and odd shape render it a most striking object. Add to this the orange aril of its seeds, peeping so daintily through the half open suture of the pod after the latter has been touched by one or two keen frosts, and we have a combination and a contrast most pleasing to the eye.

Of the fifty or more species of birds which pass the cold season in Indiana, the little winter wren, in his russet coat, is the smallest and, in **The Winter Wren.** habits, one of the most peculiar. Wherever you see him, be it on the ground, in a fence corner, or in a pile of brush or rails, he is continually on the go, flitting hither and thither, in and out of the cracks of the fence and from top to bottom of the brush pile; so that if you are a collector and want his skin you have to take him on the wing or not at all. It was, therefore, with a feeling of

delight that I heard, as I strolled along on this bright
October day, his merry "*che-che—che-che-che*," and saw
him as he ran up and down the side of a stump in
search of insects. He saw me also and seemed to
know that it was Sunday, and, therefore, he need
have no fear; for, instead of flitting away as is his
usual custom, he came nearer and nearer, seeking, as
it were, to gain my friendship, until at last I could
have touched him with my hand and even did reach
it forth, thinking that, like Thoreau's sparrow, he might
light thereon. But some spirit must have whispered
to him of his three dead kinsmen whose skins form
part of my collection; for no sooner did I stretch
forth my hand than he was up and away like a flash.
Ah, my little feathered friend, thou needst not thus
so suddenly have left my presence, for I had no
thoughts of murder in my heart, but simply wished
to bid thee welcome to thy winter's home!

I rested for a time beneath a tall white oak and
watched the falling leaves as, obedient to the great
force of gravitation, they drifted slowly towards the
center of the earth. Who can tell when a leaf breaks
its hold upon the parent tree where its resting place
will be? It comes fluttering down on account of the
broad expanse which its light weight
A Falling Leaf. presents to the resistance of the atmosphere. It is borne, now in this direction, now in
that, by the eddying gusts until at last it rests upon
a pile of a hundred others, perhaps fifty, perhaps
five hundred feet from a perpendicular from where
it started. By the action of water and oxygen, it
and its companions are soon changed into inorganic

substances and become a part of the earth's mold.
Possibly they have been a part of it hundreds, aye,
thousands of times before; for who knows what
varied forms the carbon, hydrogen and oxygen,
now locked up in the cells of the leaf, have in
the past helped to produce? Of what plant, what
animal, what man have they formed a part? But
however varied the object, which in by-gone ages
they have helped to form, it has in time fallen again
to the earth, been disintegrated, and again, by the
action of the energy of sunlight, the elements com-
posing it have been rebuilt into a new organic body
which has aided the onward march of our common
mother and fitted her better for the abode of man.

Thus we see in a falling leaf, as it were, a simili-
tude of our own lives, as we, tearing ourselves loose
from parents and home after having been nourished
to the ripeness of manhood, are borne hither and
thither by the blasts and eddyings of fate and of the
great society in which we mingle, until at last we,
too, find a resting place in the earth and yield back
to her the elements which are her own.

While musing thus over the falling leaf, the Indian
summer day, perfect as it was, came to a close. It
was, let us hope, the first of many yet to be this autumn. For on such days we enjoy the smile of nature
—tender and beautiful—her last before she dons her
seeming shroud for winter wear.

> "So we shall find—our summer being past,
> And hoar frost with us—for a little breath
> So fair a country, such a genial air,
> And shall forget our woes, and unaware
> Step over the borderland of death."

KATYDIDS AND THEIR KIN, OR THE ORTHOPTERA OF INDIANA.

The word "*Orthoptera*" means "straight-winged." It is a name given to an order or group of insects, which comprises the katydids, grasshoppers, crickets, cockroaches, walking-sticks, etc. The members of the order may be known by their *biting* mouth parts, and their indirect metamorphosis; the young when hatched being wingless, but of the same form as the parent; the wings developing gradually and appearing of full size after the skin has been shed for the fifth time, when they are membranous and, in the typical forms, laid straight along the back. In number of species the order Orthoptera is a small one, but about 850 having been described from the United States. Of these I have taken, personally, 127 in Indiana, divided among six families, as follows: *Blattidæ* or cockroaches, nine; *Phasmidæ* or walking-sticks, one; *Mantidæ* or rear-horses, two; *Locustidæ* or katydids and green grasshoppers, 39; *Gryllidæ* or crickets, 25; *Acrididæ* or short-horned grasshoppers, 51. Taking up briefly, each of these families, let us note the characters which distinguish its members, and give a few facts concerning the habits and life histories of the more common and familiar examples of each which are found in Indiana.

BLATTIDÆ.

From the other families of Orthoptera the *Blattidæ*, commonly known as cockroaches, may be known by their depressed oval form; their nearly horizontal head, which is bent downward and almost concealed by the broad chest or pro-thorax; their slender legs of equal length and size; their five-jointed tarsi or feet, and by the absence of either ovipositor or forceps-like appendages at the end of the abdomen.

The rings of the abdomen overlap each other and are capable of great extension and depression, so that these insects seem to be pre-eminently fitted for living in the narrow crevices and cracks which they inhabit. The legs are of peculiar structure, in that they are long and more or less flattened, thus enabling the cockroaches to run with surprising swiftness; so that the family has been placed by some writers in a separate sub-order, the *Cursoria* or runners. The wing covers or outer wings are leathery, translucent, and, when developed, overlap when at rest; while the wings never exceed the wing covers in length, and in some cases are rudimentary or even wanting.

From the other *Orthoptera* the cockroaches differ widely in their habits of oviposition, as the eggs are not laid one at a time, but all at once, in a peculiar capsule or egg-case, called an oötheca.* These capsules vary in the different species, as regards the size, shape, and number of eggs they contain, but they are all similar in structure. Each one is divided lengthwise by a membranous partition into two cells. Within

---

*See Fig. 44 for illustration of oötheca of Croton bug.

each of these cells is a single row of cylindrical pouches, somewhat similar in appearance to those of a cartridge belt, and within each pouch is an egg. The female cockroach often runs about for several days with an egg case protruding from the abdomen, but finally drops it in a suitable place, and from it the young in time emerge. While this method of oviposition is the one practiced by all the species of common occurrence in the United States, there seem to be exceptions to it, as Dr. C. V. Riley a few years ago recorded the fact of an introduced tropical cockroach which produced the young alive.

All young cockroaches resemble the parents in form, but are wholly wingless, the wings not appearing until after the fifth or last moult. The young are often mistaken for mature individuals by persons who have not made a careful study of the life history of the insects; and those of one or two well known and common forms have, in the past, even been described or figured as distinct, wingless species by some of the leading entomologists of the country.

Although abundantly represented in individuals, the number of species of cockroaches inhabiting the Eastern United States is comparatively few, but about twenty having been recorded. Of these, nine, representing five different genera, are known to occur in Indiana. Of the nine, seven are indigenous or native species, the other two having been introduced from the Old World.

In this connection I shall consider the habits of four of these insects, the first of which is the Oriental or black roach, *Periplaneta orientalis* (L.). This species is

dark mahogany-brown in color, and full grown specimens are about an inch in length.   The outer wings of
the female are only about one-fifth of an inch long,
while those of the male are more than half an inch and

**The Oriental
or
Black Roach.**

cover three-fourths or more of the
abdomen.   As its name indicates, it is
a native of Asia, but has been carried
from one country to another by ship-
ping.   It delights in filth and darkness, and hence in
the holds of vessels, the cellars and basements of tene-

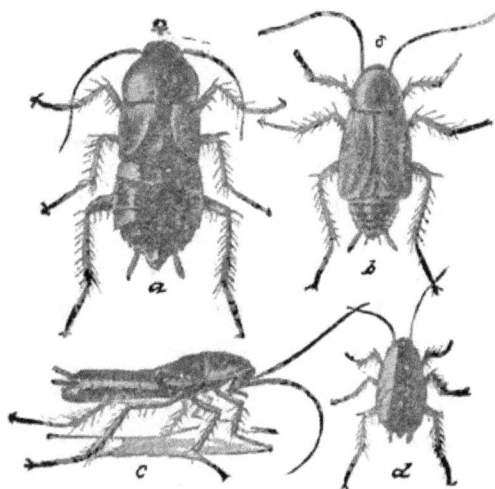

Fig. 41—Oriental or Black Roach.
(a, female; b, male; c, side view of female; d, half-grown
specimen. After Howard.)

ment houses, and in all damp, dirty places it swarms
by thousands, undoubtedly doing much good as a
scavenger, but infinitely more harm on account of its
omnivorous and insatiable appetite.   Like most other

members of the family, it feeds mainly at night, appearing to detest and avoid the light, as one can readily prove by taking a lighted lamp suddenly into its haunts, when a hurried scrambling will take place towards its daylight retreats, and but a few moments will elapse before the last of the busy marauders will have disappeared.

This is probably the most carnivorous of all our cockroaches, though, like most others, it is fond of starchy food. It is known to feed upon meat, cheese, woolen clothes, and even old leather, and is said to be especially fond of the festive bed-bug, *Acanthia lectularia* L., which soon disappears from a house infested with the Oriental roach.

Its eggs are sixteen in number, and the large horny capsule or oötheca in which they are packed is carried about by the mother for a week or longer when she drops it in a warm and sheltered place. Along one side of the capsule, which resembles in form and color a diminutive seed of the papaw, *Asimina triloba* Dunal, is a seam where the two edges are cemented closely together. When the young are hatched they excrete a liquid which dissolves the cement and enables them to escape without assistance, leaving their infantile receptacle as entire as it was before they quitted it.

In Indiana the Oriental roach is found in all the larger towns and cities, and is one of the most noisome and disagreeable insects with which certain classes of their inhabitants have to contend. It seldom occurs in houses in thinly settled localities, and never, as far as my observation goes, beneath the bark of logs and stumps.

The American cockroach, *Periplaneta americana* (L.), is, as its name implies, a native of this country, but, like the Oriental roach, it has spread to the four corners of the earth. It is by far the largest species found in the State, the male measuring 1⅗ inches from head to tip of wings, the latter in both sexes reaching beyond the end of the abdomen. The general color is reddish-brown, the top of the thorax being margined rather broadly with yellow. In Indiana it seems to be of rather limited distribution, as I know of its occurrence in but two counties, Putnam and Marion. It occurs in numbers in some of the leading hotels of Indianapolis, but usually confines itself to the basement and first floor, and appears to be much more cleanly in its choice of an abiding place than does the closely allied Oriental roach.

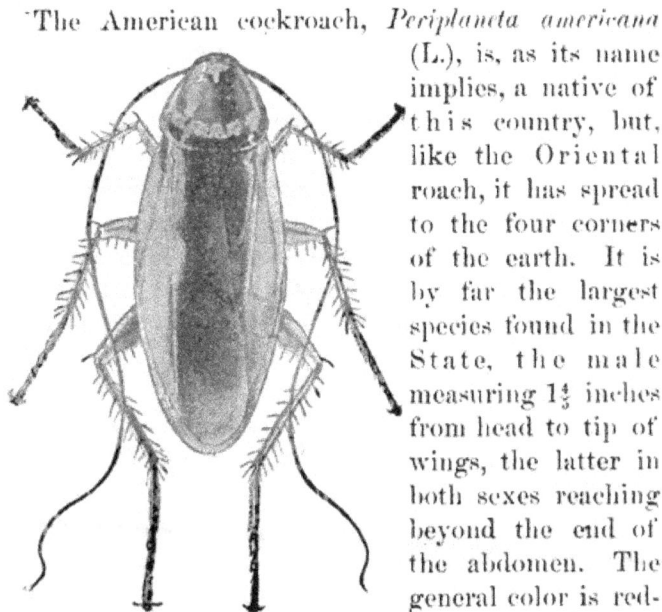

Fig. 42—American Cockroach.
(Enlarged one-third. After Howard.)

The American Cockroach.

The Pennsylvania cockroach, *Ischnoptera pennsyl-ranica* (DeGeer), is also a native species, and is the

most common roach in the State, being found everywhere beneath the loose bark of logs and old stumps.

The Pennsylvania Cockroach. It is usually seen in the wingless
stages, the mature individuals being
common only from May to October.
The half grown young are of a shining,
dark brown color, the dorsal surface of the thoracic
segments often lighter. The wings of adult specimens
are long and narrow, extending in both sexes much
beyond the tip of the abdo-
men. The total length is more
than an inch and the color is
reddish-brown with a whitish
stripe on the margins of wings
and thorax.

As mature specimens are attracted by light, country houses
are often badly infested with
them; and where food is scarce
the wall paper is sometimes
much injured for the sake of

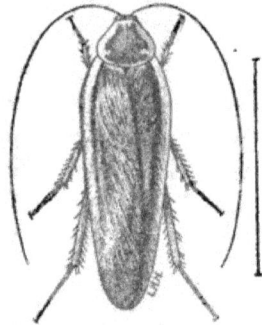

Fig. 43—Pennsylvania Cockroach. (After Lugger.)

the paste beneath. What the hordes of young which
dwell under the bark of logs live upon is a question
as yet unsettled, but the larvæ of other insects undoubtedly form a portion of their food, as in two
instances I have found them feeding upon the dead
grubs of a Tenebrio beetle; while living, as well
as decaying, vegetable matter probably forms the
other portion. The mating of the adults mostly
occurs in late summer and early autumn, the newly
hatched young being most abundant from mid-
September until December. The young in various

stages of growth survive the winter in the places
mentioned, they being the most common insects noted
in the woods at that season.   Cold has seemingly but
little effect on them, as they scramble away almost as
hurriedly when their protective shelter of bark is
removed on a day in mid-January with the mercury
at zero, as they do in June, when it registers a hundred in the shade.

The empty egg cases of the Pennsylvania roach are
very common objects beneath the loose bark of logs
and especially beneath the long flakes of the shell-
bark hickory.   They are chestnut brown in color,
about $\frac{1}{3}$ x $\frac{1}{8}$ inches in size, and are much less flattened
than those of the Croton bug, described below.   The
dorsal or entire edge is slightly curved or bent
inwards, after the fashion of a small bean.   The
young, after hatching, evidently escape in the same
manner as do those of the Oriental cockroach, as no
break is visible in the empty capsule.

The last of the four cockroaches to be considered is
the "Croton bug," *Phyllodromia germanica* (L.), so
called because it made its appearance in New York
City in numbers about the time the Croton aqueduct
was completed.   It is a native of Central Europe, but
like the Oriental roach has become cosmopolitan.

This is one of the smallest of the nine cockroaches
known to occur in the State; the total length being
$\frac{1}{2}$ inch or less.   The general color is a
The Croton Bug. light brownish-yellow, the females
often darker.   The thorax has two
dark brown bands, enclosing a yellowish stripe.

The egg case of the Croton bug is very light brown,

a little over twice as long as broad, with the sides
somewhat flattened and the edges parallel. Within
it the eggs, thirty-six in number, are arranged in the
usual two rows. It is carried about by the mother
roach for several days with from half to three-fourths
of its length protruding from the abdomen, and when
dropped in a favorable place the young evidently
very soon emerge from it; for in a bottle in which a
female with protruding oötheca was placed at eleven
o'clock P. M. the young were found to have emerged
on the following morning at eight.

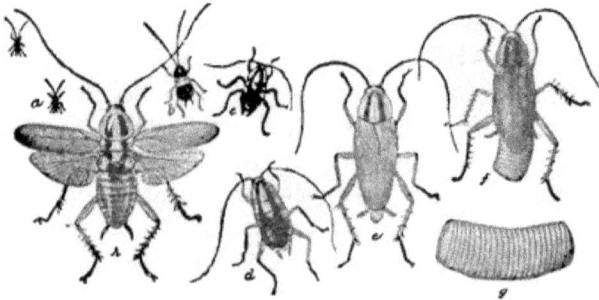

Fig. 44—Croton Bug.

(*a*, first stage; *b*, second stage; *c*, third stage; *d*, fourth stage; *e*, adult; *f*, adult
female with egg case; *g*, egg case enlarged; *h*, adult with wings spread. After
Howard.)

The Croton bug seldom, if ever, occurs in numbers
in the country, but is one of the worst insect pests
with which the inhabitants of the larger cities of Indiana have to deal. It is the most fecund of all the
roaches, and the seasons of mating and hatching of
the young are, perhaps, more irregular than in any
other species. Adult forms are evidently to be found
at all seasons of the year, as I have taken them in

December, April and October. It is not so much a lover of filthy surroundings as is the Oriental roach, and hence frequents much more often than that species the dwellings of the better class of people. It delights in warm, moist places, and is especially abundant and destructive in buildings which are heated by steam. Where it once obtains a foothold and the surroundings of temperature and food supply are favorable, it is almost impossible to eradicate, as its small, flattened form enables it to hide and breed in cracks and crevices, which none of the larger roaches can enter.

Like many other omnivorous animals, Croton bugs find in wheaten flour a food substance which is rich in nutrition and easily digested, and so prefer wheat breads and starchy materials to all other foods. On account of this liking they often do much harm to cloth-bound books by gnawing their covers in search of the paste beneath. They also seem to have a peculiar liking for paints of various kinds, and in the office of the U. S. Coast and Geodetic Survey at Washington, have done much damage by eating off the red and blue paints from the drawings of important maps. On one occasion they made a raid on a box of water colors, where they devoured the cakes of paint, vermilion, cobalt and umber, alike; and the only vestiges left were the excrements in the form of small pellets of various colors in the bottom of the box.

In giving a remedy for this and other species of cockroaches which frequent houses, I can not do better than quote from Dr. Riley's excellent article in "Insect Life." He says: "Without condemning other useful measures or remedies like borax, I would repeat that in the free and persistent use of California Buhach, or some other fresh and reliable brand of Pyrethrum or Persian Insect Powder, we have the most satisfactory means of dealing with these roaches.

"Just before nightfall go into the infested rooms and puff into all crevices, under base-boards, into drawers and cracks of old furniture—in fact, wherever there is a crack—and in the morning the floor will be covered with dead and dying or demoralized and paralyzed roaches, which may easily be swept up or otherwise collected and burned.

"With cleanliness and persistency in these methods the pest may be substantially driven out of a house, and should never be allowed to get full possession by immigrants from without.

"For no other insects have so many quack remedies been urged and are so many newspaper remedies published. Many of them have their good points, but the majority are worthless. In fact, rather than put faith in half of those which have been published, it would be better to rely on the recipe which is current among the Mexicans:

"'To Get Rid of Cockroaches.—Catch three and put them in a bottle, and so carry them to where two roads cross. Here hold the bottle upside down, and as they fall out repeat aloud three *credos*. Then all the cockroaches in the house from which these three came will go away.'"

PHASMIDÆ.

Belonging to the family *Phasmida*, there occurs in Indiana, as far as known, but one species, the walking-stick, *Diapheromera femorata* (Say). This insect furnishes

The Walking-stick. a most excellent example of adaptation for the purpose of protection. It is wingless and possesses a long, cylindrical body, resembling a slender stick with the bark on it. It moves very slowly, and has a habit of remaining motionless and apparently dead for a considerable length of time. On such occasions it usually stretches itself out from a twig, with its front legs and antennæ extended, and then can scarcely be distinguished from a prolongation or branch of the twig. Many people who see them thus for the first time and afterwards watch them moving slowly away, can scarcely be persuaded that they are not real twigs, gifted in some mysterious manner with life and motion.

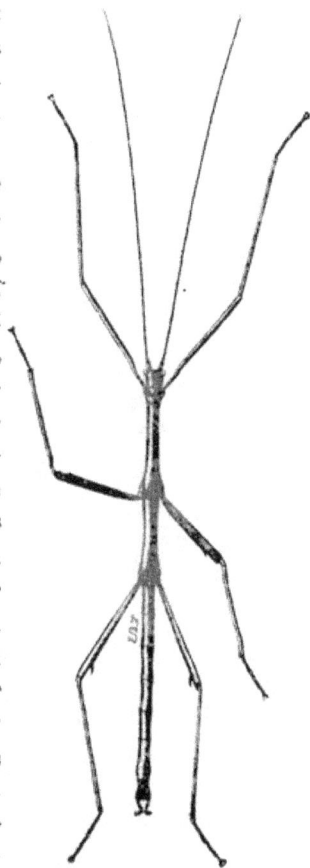

Fig. 45—Walking-stick.
(After Lugger.)

The walking-stick feeds, during its entire life, upon the foliage of various trees, being especially fond of the leaves of the wild cherry. The eggs are dropped upon the ground by the mother insect, who takes no farther notice of them. The young, when hatched, trust to chance and their peculiar shape to escape those higher forms which are ever ready to prey upon every living thing which promises them a bit of sustenance.

MANTIDÆ.

Two species of the family *Mantidæ* are found sparingly in southern Indiana. Of these the best known is the Carolina mantis, or rear-horse, one of the two

The Carolina Mantis. or three species of Orthoptera, which are in the slightest degree beneficial.

It is a rather large insect, of a greenish or gray color. When disturbed it elevates or rears the fore part of its body almost perpendicularly, fixes its large, staring eyes upon the intruder and turns its head sideways in a very ludicrous manner, so as to follow, if possible, every movement of its supposed enemy. If a small object, as a blade of grass, be then thrust towards it, it will strike out vigorously with its saber-like fore feet, or else retreat to what it considers a safe distance.

Being a rapacious insect, its fore limbs have, in time, become peculiarly adapted for grasping organs. The tibiæ are robust and armed with a double row of spines; the tarsi or feet are short, spiny and curved so as to fit into the under side of the tibiæ like the blade of a clasp knife when closed. When in pursuit

14

of its prey the mantis moves almost imperceptibly along, stealing towards its victim like a cat approaching a mouse. When sufficiently near, the fore leg is suddenly extended to its full length and the unlucky insect is immediately caught and impaled by the spines between the tibiæ and tarsi, carried to the mouth and deliberately eaten piecemeal while yet alive and struggling to escape.

The Carolina mantis is occasionally found as far north as Indianapolis, where it reaches maturity about September 1st. When the two sexes are cap-

Fig. 46—Carolina Mantis.

tured and placed together the female soon begins to feed upon her liege lord, and finally devours all portions of him which are in the least degree digestible. The eggs are grouped together in agglutinated masses of 40 or more and covered with a coarse web of silk, the top of the mass then appearing as if its component parts were braided together. In this manner the species survives the winter and in May, when insect life begins to abound, the young emerge and use their prominent staring eyes to good advantage in seeking plant lice and other minute forms which furnish them their first of many meals.

LOCUSTIDÆ.

The family of Orthoptera known as the *Locustidæ* comprises those insects commonly called katydids, green grasshoppers, and stone or camel crickets. The distinguishing characters of the members of this family are the long, slender, tapering, many-jointed antennæ; the almost universal absence of simple eyes; the four-jointed tarsi or feet; and the sword-shaped ovipositor of the females. The outer wings, when present, slope obliquely downwards, instead of being bent abruptly, as in the cricket family; and in most cases the wings are longer than the wing covers.

The stridulating or musical organ of the males is quite similar in structure to that of the male cricket, being found at the base of the overlapping dorsal surface of the wing covers and usually consisting of a transparent membrane, of a more or less rounded form, which is crossed by a prominent curved vein bearing on the under side a single row of minute file-like teeth. In stridulating, the wing covers are moved apart and then shuffled together again, when these teeth are rubbed over a vein on the upper surface of the other wing cover, producing the familiar, so-called "katydid" sound. Each of the different species makes a distinct call or note of its own, and many of them have two calls, one which they use by night and the other by day. Any one who will pay close attention to these different calls can soon learn to distinguish each species by its note as readily as the ornithologist can recognize different species of birds in the same manner. The ear of these insects,

when present, is also similar in structure and position
to that of the cricket's, being an oblong or oval cavity covered with a transparent or whitish membrane,
and situated on the front leg near the basal end of
the tibia.

The young of the *Locustidæ*, like those of the other
families of the order, when hatched from the egg
resemble the adults in form but are wholly wingless.
As they increase in size they moult or shed the skin
five times, the wings each time becoming more appar-
ent, until after the fifth moult, when they appear fully
developed, and the insect is mature or full grown,
never increasing in size thereafter. Throughout their
entire lives they are active, greedy feeders, mostly
herbivorous in habit; and, where present in numbers,
necessarily do much harm to growing vegetation.
Thirty-nine species of the family are known to occur
in Indiana.

Popularly speaking, we may divide the members
of the family into three groups, the katydids, green
grasshoppers and stone or camel crickets. Taking
these up in the order mentioned, we find that the
"katydids," nine species of which have, up to the
present, been recorded from the State, are the most
arboreal of all the *Locustidæ*. The great majority of
them pass their entire lives on shrubs and trees,
where they feed upon the leaves and
The
Katydids.   tender twigs, and, when present in
numbers, often do excessive injury.
The color and form of their wings serve admirably to
protect them against their worst foes, the birds; and
as they live a solitary life, *i. e.*, do not flock together

in numbers as do the green grasshoppers, they are but seldom noticed by man. Their love calls or songs, however, make the welkin ring at night from mid-August until after heavy frost, and though but one of the nine species found in the State makes a note in any way resembling the syllables "katy-did, she-did," yet all are accredited with this sound by the casual observer, and hence the common name usually given to the members of this sub-family. Their call is seldom made by day for the obvious reason that it might attract the attention of the birds and so lead to the destruction of the insect. As twilight approaches, however, the male of each species begins his peculiar note, which is kept up, with little or no intermission, until the approach of day warns him that his feathered enemies will soon be on the alert, and that silence will be, for a time, the best policy to pursue.

From the other *Locustidæ* the katy-dids differ widely in their habits of oviposition, the eggs not being deposited on the earth or in twigs, but are usually glued fast in double rows to the outer surface of slender twigs or on the edges of leaves. The eggs of the most common species appear like flattened hemp seeds, and usually over-lap one another in the row in which they are placed. On account of this method of oviposition, the ovipositors

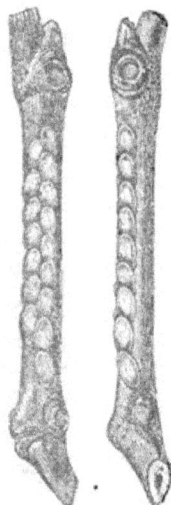

Fig. 47—Eggs of Angular-winged Katydid.

of the katydids are broader, more curved and more

obtuse at the end than in the other sub-families, whose members oviposit in the earth or in the stems of grass.

The broad-winged or "true katydid," *Cyrtophyllus concavus* (Harris), is found in considerable numbers

The Broad-winged Katydid. throughout the State, but is much more commonly heard than seen, as it dwells singly or in pairs in the densest foliage which it can find, such as the tops of shade trees and the entwining vines of the

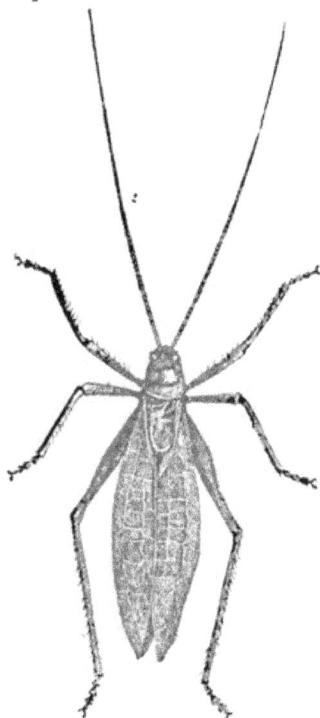

Fig. 48—Broad-winged Katydid.

(Male. After Harris.)

grape arbor. In central Indiana it reaches maturity as early as July 20; and is more domestic in its habits than any other species of the "katydid" group, frequenting, for the most part, the shrubbery of yards, orchards and the trees along fence rows, being seldom heard in extensive wooded tracts. Its note is the loudest made by any member of the family, the male having the musical organ larger and better developed than in any other. The call is almost always begun soon after dusk with a single note uttered at intervals of about five seconds for a half dozen

or more times. This preliminary note gives the
listener the impression that the musician is tuning
his instrument preparatory to the well known double
call which is soon begun and kept up almost continuously from dark till dawn.

Of this call, Mr. S. H. Scudder has written: "The
note of the true katydid, which sounds like *Xr*, has a
shocking lack of melody; the poets who have sung
its praises, must have heard it at the distance that
lends enchantment. In close proximity, the sound is
excessively rasping and grating—louder and coarser
than I have ever heard from any of the Locustarians
in America or in Europe, and the Locustarians are
the noisiest of all *Orthoptera*. Since these creatures
are abundant wherever they occur, the noise produced
by them, on an evening especially favorable to their
song, is most discordant. Usually, the notes are two
in number, rapidly repeated at short intervals. Perhaps nine out of ten will ordinarily give this number,
but occasionally a stubborn insect persists in sounding the triple note—'katy-she-did'—and as katydids
appear desirous of defiantly answering their neighbors in the same measure, the proximity of a treble-
voiced songster demoralizes a whole neighborhood,
and a curious medley results. Notes from some individuals may then be heard all the while, scarcely a
moment's time intervening between their stridula-
tions—some nearer, others at a greater distance—so
that the air is filled by these noisy troubadours with
an indescribably confused and grating clatter."

The "angular-winged katydid," *Microcentrum lauri-
folium* (L.), is another species which in the country is

commonly called "the katydid," and the note of *Cyrtophyllus concavus* is usually attributed to it but its

The Angular-winged Katydid. true note may be represented "by the syllable '*tie*,' repeated from eight to twenty times at the rate of about four to the second." It is evidently attracted by light, being often found in the gutters beneath the electric lights in the larger cities and towns. It occurs, probably, throughout the State, but is more common southward and is nowhere found in sufficient numbers to be injurious. The eggs are laid on twigs which have been previously roughened with the jaws and otherwise prepared for a place of deposit. The two rows are contiguous and the eggs of one alternate with those of the other. Those of the same row overlap about one-fourth of their length. They are of a grayish brown color, long oval in shape, very flat, and measure 5.5 x 3 mm. They are usually deposited in September, hatch the following May, and the young, in central Indiana, reach maturity during the first half of August.

Fig. 49.—Angular-winged Katydid. (Male. After Riley.)

The green grasshoppers are those slender-bodied *Locustidæ*, with long, tapering antennæ, which are so

common in summer and early autumn in damp meadows and prairies, and along the margins of streams, ditches and ponds.
They are mostly terrestrial in their
habits, but one or two of the larger ones ever being
found in trees.

The Green Grasshoppers.

Of this group 21 species have been found in Indiana, four of which, called "cone-headed grasshoppers,"
are more than twice as large as the others and have
the vertex or top of the head prolonged forwards and
upwards into a prominent cone. The outer wings are
long and slender, and the ovipositor is oftentimes of
excessive length. The total length of the females is
between one and a half and two inches, that of the
males averaging about a third less.

These insects seem to "possess more intelligence than
is usual among the Orthoptera, and they are about
the most difficult of the order to approach. In escaping they usually slip or fall into the grass instead of
jumping or flying; but they seem to fully understand
that they are very well protected by their color and
form. If approached very cautiously they often remain quite still upon the stem of grass upon which
you have surprised them, with the usually well
founded expectation that you will not be able to distinguish them from the green herbage around. If
they think it worth while to make some active movement to escape they will frequently slip around on
the other side of the stem and walk down it to
the ground or off upon another plant. Unlike most
Orthoptera they do not use their front legs in holding to the mouth the thing upon which they feed.

Instead of biting they seem to wrench or tear away pieces from the stems or leaves."

The females of the cone-heads deposit their eggs between the stems and root leaves of coarse grasses and sedges. The young are hatched in May and reach maturity about the 5th of August. The notes of the male vary much according to species, that of the more common "sword-bearer," *Conocephalus ensiger* Harris, being similar to the syllables "*ik-ik-ik*," as if sharpening a saw, this sound enlivening low

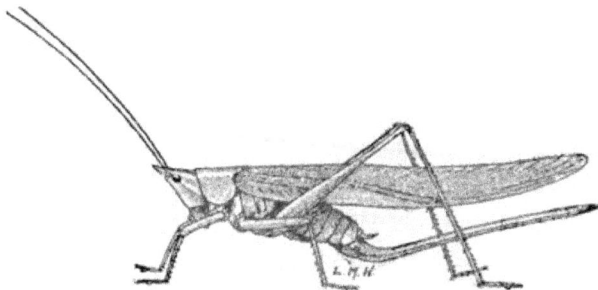

Fig. 50—Sword-bearer.
(Female. After Lugger.)

bushes, and particularly the corn patch, as it seems to especially delight in perching near the top of a cornstalk and there giving forth its rather impulsive song.

The other members of the green-grasshopper group, 17 in number known to occur in the State, seldom exceed three-fourths of an inch in length. The color of their bodies corresponds closely to that of the stems and leaves of the sedges and grasses among which they dwell, and so protects them from the sight of the few birds which frequent a like locality. Their songs, produced in the same manner as those of their larger cousins, the katydids, are as frequent by day as by night, but are usually soft and low in comparison with those of the former. Their day song differs from that of the night, and, "it is curious to observe these little creatures suddenly changing from the day to the night song at the mere passing of a cloud, and returning to the old note when the sky is clear. By imitating the two songs in the daytime, the grasshopper can be made to represent either at will; at night they have but one note."

Fig. 51—Lance-tailed Grasshopper.
Xiphidium attenuatum Scudder.

The eggs of these smaller green grasshoppers are deposited within the stems or root leaves of grass, the pith of twigs, or sometimes in the turnip-shaped galls so common on certain species of willow. The ovipositor being thus used as a piercer, has in time developed into a slender and sharp-pointed instrument which is but little curved and is frequently of excessive length, in some species being over twice as long as the remainder of the body.

Eight of the seventeen belong to the genus *Xiphidium*, meaning "sword-bearer," which includes the smallest and most slender-bodied of the winged *Locustidæ*.

A very common member of this genus in western Indiana is *Xiphidium nemorale* Scudder. It reaches maturity about August 15th and from then until after heavy frosts may be found in numbers along the borders of dry, upland woods, fence rows and roadsides, where it delights to rest on the low shrubs, blackberry bushes, or coarse weeds usually growing in such localities. On the sunny afternoons of mid-autumn it is especially abundant on the lower parts of the rail and board fences, the male uttering his faint and monotonous love call—a sort of *ch-e-e-e-e—ch-e-e-e-e*, continuously repeated—the female but a short distance away, a motionless, patient, and apparently attentive listener.

The remaining nine members of the group belong to the genus *Orchelimum*, the literal meaning of which is, "I dance in the meadows." The name is a most appropriate one, for low, moist meadows everywhere swarm with these insects from July to November; and though waltzes and quadrilles are probably not indulged in, yet the music and song, the wooing and love making, which are the natural accompaniments of those amusements, are ever present, and make the short season of mature life of the participants a seemingly happy one.

Among these the "common meadow grasshopper," *Orchelimum vulgare* Harris, is probably the most abundant member of the family *Locustidæ* found in Indiana. It begins to reach maturity in the central part of the State about July 20th, and more frequently

than any other of our species of *Orchelimum* is found
in upland localities, along fence rows, and in clover
and timothy mead-
ows. In early au-
tumn it is very fond
of resting on the
leaves and stems of
the iron-weed so
common in many
blue-grass pastures.

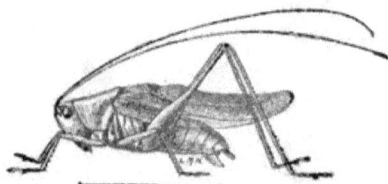

Fig. 52—Common Meadow Grasshopper.
(Male. After Lugger.)

This green grasshopper seems to be somewhat carnivorous in habit, as on two occasions I have discovered it
feeding upon the bodies of small moths which in some
way it had managed to capture. The note of the male
is the familiar *zip-zip-zip-zip—ze-e-e-e*—the first part
being repeated about four times, usually about twice
a second; the *ze-e-e-e* continuing from two or three to
twenty or more seconds.

> "The poetry of earth is never dead:
>   When all the birds are faint with the hot sun,
>   And hide in cooling trees, a voice will run
> From hedge to hedge about the new mown mead;
> That is the grasshopper's—he takes the lead
>   In summer luxury, he has never done
>   With his delights; for when tired out with fun
> He rests at ease beneath some pleasant weed."

The "stone or camel crickets," nine species of
which occur in Indiana, are wingless *Locustidæ* of

**The Stone or
Camel Crickets.**
medium or large size with a thick
body and arched back. They are seldom seen except by the professional
collector, as they are nocturnal in their habits, and
during the day hide beneath stones along the margins

of small woodland streams, or beneath logs and chunks in damp woods, in which places seldom less than two, nor more than three or four, are found associated together.

That they are well nigh omnivorous in their choice of food, I have determined by keeping them in con-

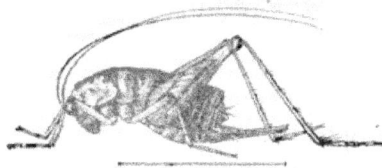

Fig. 53—Stone or Camel Cricket.
Ceuthophilus maculatus (Say).
(Female. After Lugger.)

finement, when they fed upon meat as well as upon pieces of fruit and vegetables, seemingly preferring the latter. The majority of the species evidently reach maturity and deposit their eggs in the late summer or early autumn, as the full grown insects are more common then, but have been taken as late as December 1st. The eggs, which are supposed to be laid in the earth, usually hatch in April; but some are hatched in autumn and the young live over winter (an anomaly among the Locustidæ), as I have taken them in January and February from the localities which the adults frequent in summer. Several of the species inhabit caves and are usually of much larger size, with longer antennæ and smaller compound eyes than those found above ground.

GRYLLIDÆ.

The *Gryllidæ* or crickets are, in the main, distinguished from other Orthopterous insects, by having the wing covers flat above and bent abruptly downward at the sides; the antennæ long, slender, and many jointed; the tarsi, or feet, three jointed, without pads between the claws; the ear situated on the tibia of the fore leg, and the abdomen bearing a pair of jointed cerci or stylets at the end.

The ovipositor of the female, when present, is long, usually spear-shaped, and consists, apparently, of two pieces. Each of these halves, however, when closely examined, is seen to be made up of two pieces so united as to form a groove on the inner side; so that when the two halves are fitted together, a tube is produced, down which the eggs pass to the repository in the earth or twig fitted to receive them.

Representatives of 25 species of these interesting insects have been taken in Indiana, several of which are exceedingly abundant throughout the State. Among these are two species of burrowing or "mole crickets" which rank first in size and singularity of structure. When full grown they measure from one inch to an inch and a half in length; are light brown in color and have

The Mole Crickets.

the body covered with very short hairs, giving it a soft velvety appearance. The females have no visible ovipositor, and, externally, may be separated from the males only by the difference in the veining of the uppermost of the wing covers. By their habit of bur-

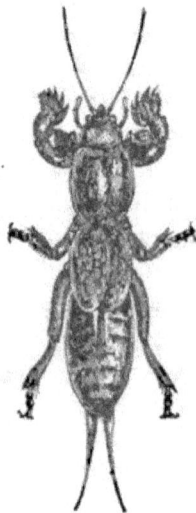

Fig. 54—Mole Cricket.

rowing beneath the soil in search of such food as the tender roots of plants, earth-worms and the larvæ of various insects, the anterior tibiæ of these crickets have, in the course of ages, become so modified in structure as to closely resemble the front feet of the common mole, whence the generic name, *Gryllotalpa*, from "gryllus," a cricket, and "talpa," a mole. Moreover, the compound eyes have become very much aborted, being not more than one-eighth the size of those of the common field cricket, *Gryllus abbreviatus* Serv.; and, as the insect crawls rather than leaps, the hind femora are but little enlarged.

The mole crickets are found in all parts of Indiana, though nowhere in great abundance. Their eggs are laid in under-ground chambers in masses of from forty to sixty, and the young are about three years in reaching maturity. On this account, where they exist in numbers, they are very destructive, feeding, as they do during that time, mainly upon the tender roots of various plants. It is therefore fortunate that with us these crickets are not more common than they are. In the moist mud and sand along the margins of the smaller streams and ponds their runs or burrows, exactly like those of a mole, though much smaller, can in late summer and early autumn be seen by those interested enough to search for them. These runs usually end beneath a stone or small stick, but the insects are very seldom seen, as they are nocturnal, forming their burrows by night, and scarcely ever emerging from beneath the ground.

The note of the male is a sharp di-syllabic chirp, continuously repeated and loud enough to be heard several rods away. It is usually attributed, by those who have given little attention to insect sounds, to the field crickets or to some of the smaller frogs. They are very difficult to locate by this note, and I have on several occasions approached cautiously on hands and knees, a certain spot and have remained silent for some minutes while the chirping went on, apparently beneath my very eyes; yet when the supposed exact position of the chirper was determined and a quick movement was made to unearth him he could not be found. Indeed, it is only by chance, as by the sudden turning over of a log in a soft, mucky place that a person can happen upon one of them unawares. Even then quick movement is necessary to capture him before he scrambles into the open mouth of one of the deep burrows which he has ever in readiness.

Probably the best known crickets in the State are the "field crickets"—those dark-colored, thick-bodied species, mature specimens of which are so abundant from late summer until after heavy frosts, beneath logs, boards, stones, and especially beneath rails in the corners of the old-fashioned and rapidly disappearing rail fences. The eggs of some of the

The Field Crickets.

field crickets are laid in the ground in late autumn and hatch the following May. Those of at least two species are, however, laid in late summer or early autumn, and hatch before frost, the half grown young being found in numbers throughout the winter beneath logs and chunks. On cold days they are usually found in a dormant condition, each one at the bottom of a cone-shaped cavity

15

which it has formed for itself, and which is very similar to the pit made in loose sand by the larva of the ant lion, *Myrmeleon obsoletus* Say.

The most common of the five species occurring with us is the short-winged field cricket, *Gryllus abbreviatus* Serv., which is nocturnal, omnivorous, and a cannibal. Avoiding the light of day he ventures forth, as

Fig. 55—Field Cricket.
(Female.)

soon as darkness has fallen, in search of food, and all appears to be fish which comes to his net. Of fruit, vegetables, grass and carrion he seems equally fond, and does not

hesitate to prey upon a weaker brother when opportunity offers. I have often surprised them feasting upon the bodies of their companions, and of about forty imprisoned together in a box, at the end of a week but six were living. The heads, wings and legs of their dead companions were all that remained to show that the weaker had succumbed to the stronger; that the fittest, and in this case the fattest, had survived in the deadly struggle for existence.

Of all the *Gryllidæ* which occur in the Northern States, the little brown "ground crickets" are the most numerous and the most social. Unlike their larger cousins, the field crickets, they do not

The Ground Crickets. wait for darkness before seeking their food; but wherever the grass has been cropped short, whether on shaded hillside or in the full glare of the noonday sun along the beaten roadway, mature specimens may be seen by hundreds during the days of early autumn. They are all of small size, being seldom more than half an inch in length. The color is a dark brown, and the bodies and legs are sparsely clothed with brown hairs.

These crickets are omnivorous, feeding upon all kinds of decaying animal matter as well as upon living vegetation. When disturbed they are very difficult to capture, making enormous leaps with their stout hind legs, no sooner striking the ground than they are up and away again, even if not pursued, until they find a leaf or other shelter beneath which to take refuge. Six species occur in Indiana and from their enormous numbers, as well as from the fact that they are constant, greedy feeders from the time the eggs hatch in the spring until laid low by the hoar frost of autumn, it follows that they must be classed among our most injurious Orthoptera, but as yet no effective means for their destruction has been discovered.

Fig. 56—Ground Cricket. *Nemobius fasciatus vittatus* Harris. (Female, twice natural size. After Lugger.)

Among the crickets occurring in Indiana is a short, thick-bodied brown form, *Apithes agitator* Uhler. It has been taken in several of the south-western counties, notably in Vigo, where the first ones discovered were on the slender twigs of some prickly-ash shrubs which grew in a damp upland woods. The place was visited a number of times and the crickets were always found, perfectly motionless, and immediately above or below one of the thorns or prickles jutting forth from the twigs. The tips of the hind femora were raised so as to project above the body, thus causing them to resemble the thorns; and the color of the insects, corresponding closely with that of the bark, made them very difficult to discover even when in especial search of them. On every clump of prickly-ash in the woods mentioned a number of specimens were secured but they could be found nowhere else thereabouts. On another occasion they were discovered about the roots of a scarlet oak, *Quercus coccinea* Wang, which grew on a sandy hillside. Here they were plentiful, and resting motionless in the depressions of the bark or beneath the leaves in the cavities formed by the roots of the tree.

Of all the males taken in both places, over thirty in number, there was not one with perfect wing covers, and, in almost every instance, the wing covers as well as the rudimentary wings were wholly absent; while every female had both pairs unharmed. I at first ascribed this wing mutilation to the males fighting among themselves, but finally discovered a female in the act of devouring the wings of a male. Why this curious habit on the part of the one sex? Possibly the females require a wing diet to requite them for their bestowed affections, or, perchance, they are a jealous set, and, having once gained the affections of a male, devour his wing covers to keep him from calling other females about him.

The tree crickets may be known from others of their kin by their slender hind legs, their narrow, elongated chest or pro-thorax, and their whitish or greenish-white color. The wing covers of the females

The Tree Crickets. are wrapped closely about the body, while those of the male are much firmer in texture, broadly spread out, and very transparent; causing such a difference of appearance between the two sexes that tyro collectors often take them for widely different insects. Of the six species known to occur in Indiana the snowy tree cricket, (*Ecanthus niveus* DeGeer, is the most common and the best known. Both sexes of this species are in color ivory white, more or less tinged with a delicate green, especially in the

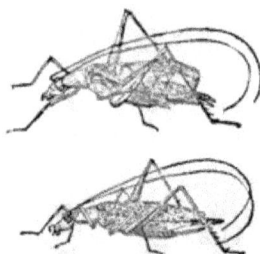

Fig. 57—Snowy Tree Crickets. (Male and female.)

female. The top of the head and basal joint of antennæ are usually suffused with ochre yellow, while on the lower face of each of the two basal joints of the antennæ is a small black spot. The ovipositor of the female is short, perfectly straight and usually tipped with black.

The snowy tree cricket is very common throughout the State, and mature specimens are to be found in numbers about grape vines, shrubbery, etc., from August 1st until November. The females appear more plentiful than the males, the latter being more often heard than seen. During the day they keep themselves hidden among the foliage and flowers of various plants, but as night approaches they come forth and the male begins his incessant, shrill, chirping note—the well known *t-r-r—r-c-c*; *t-r-r—r-c-c*, repeated without pause or variation about seventy times a minute.

The females of the snowy cricket do much harm by ovipositing in the tender canes or shoots of various plants, as the raspberry, grape, plum, peach, etc.; no less than 321 eggs, by actual count, having been found in a raspberry cane 22 inches in length. The eggs are laid in autumn and at first the injury is shown only by a slight roughness of the bark, but afterwards the cane or branch frequently dies above the puncture, or is so much injured as to be broken off by the first high wind. If the injured and broken canes containing the eggs be collected and burned in early spring the number of insects for that season will be materially lessened.

This injurious habit is partly, if not wholly, offset by the carnivorous habits of the crickets, as the young, which are hatched in June, feed for some time upon the various species of aphides or plant lice which infest the shrubbery they frequent. Miss Mary E. Murtfeldt, of Kirkwood, Mo., has given an interesting account of some experiments and observations concerning this habit,

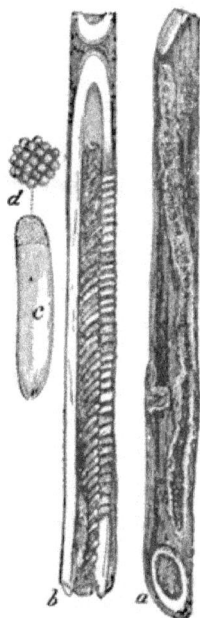

Fig. 58—Eggs of Tree Cricket in raspberry cane.

a, Cane, showing puncture; *b*, cane split to show eggs; *c*, egg enlarged.

from which the following extract is taken: "Some leaves of plum infested with a delicate species of yellow aphis were put into a jar with the young of *Œcanthus niveus*, but attracted no immediate attention. As twilight deepened, however, the crickets awakened to greater activity. By holding the jar against the light of the window, or bringing it suddenly into the lamp light, the little nocturnal hunters might be seen hurrying with a furtive, darting movement over the leaves and stems, the head bent down, the antennæ stretched forward, and every sense apparently on the alert. Then the aphides provided for their food would be caught up one after another with eagerness and devoured with violent action of the mouth parts, the antennæ meanwhile playing up and down in evident expression of satisfaction. Unless I had provided very liberally not an aphis would be found in the jar the next morning, and the sluggish crickets would have every appearance of plethora."

ACRIDIDÆ.

The common things about us, those which we meet in our every-day life, are usually those of which we know the least. Everybody knows a grasshopper by sight. How many can describe the salient points in its life history, can tell of the many devices which it uses to avoid its enemies, or of the many ways in which its organs have become adapted to or fitted for the life it leads? Yet any one interested in the objects of nature can soon learn these and other similar facts for himself by a little patient, personal investigation.

To the people of ancient times the "grasshopper" of to-day was the "locust," one of the seven plagues of Egypt. The scientific men of this country have long endeavored to have them called "locusts" in the United States, but the majority of people persist in calling them "grasshoppers," and give the name "locust" to those noisy insects which once every seventeen years invade our fields and forests in such countless numbers. To my mind "grasshopper" is the better and more expressive name for the insect with which we have to deal, and, though "locust" has the priority, as the wise men say, yet we shall relegate it to the shades of the past and know our subject by its most common title.

To the average observer a grasshopper is a grasshopper, nothing more—like

"The primrose on the river's brim"

was to Peter Bell. But to the naturalist, or to any person who will keep his eyes and ears open as he walks about, there are grasshoppers and grasshoppers—not individuals, but different kinds, each with interesting facts to be learned concerning it. Over 520 species inhabit the United States, 51 of which have been taken by myself within the limits of Indiana.

Most of these pass the winter in the egg stage, the eggs, in early autumn, being deposited in the earth by the mother insect in compact masses of forty to sixty each in the manner shown in the accompanying cut. About mid-April these eggs begin to hatch and the sprightly little insects, devoid of wings but otherwise like their parents, are soon seen on every hand.

Born with one earthly desire—a voracious appetite—and with one valuable possession—a pair of strong,

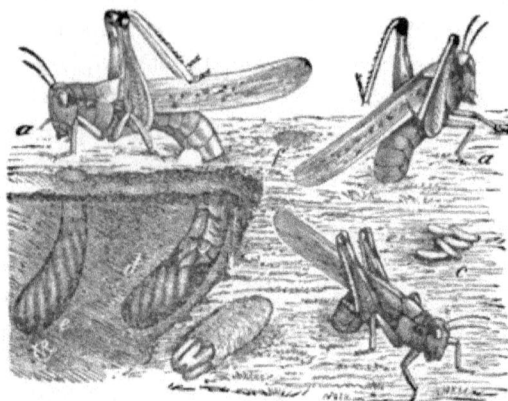

Fig. 59—Grasshopper in the act of laying eggs. (After Riley.)

broad jaws, which move in and out like the blades of a pair of scissors—the little hopper soon begins to use

The Young of Grasshoppers.

the latter to appease the former, and for twenty-four hours a day and seven days in a week, he gnaws away at the soft, green, succulent grass which surrounds him on every side. Such a procedure can have but one result. His body soon becomes too big for its surroundings. Something must give way and that something is his skin. He casts it aside with but little reluctance, however, for a new one is ready to take its place, and immediately begins to satiate his appetite once more. Five successive times his skin gets too small for his body and is cast aside. Between each of these moults the wings are growing and when the fifth skin is shed he emerges a mature and fully fledged insect.

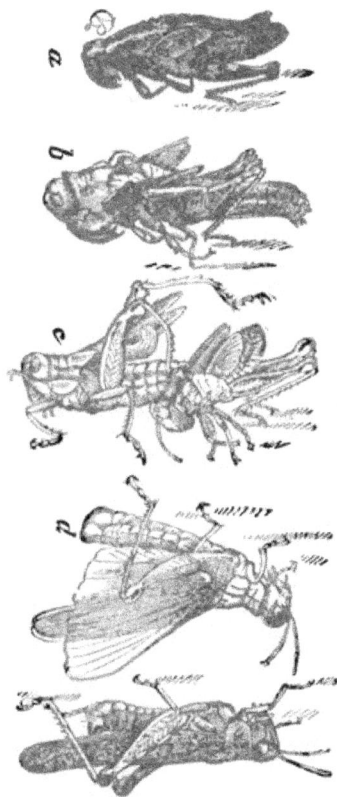

While passing through a field on one September day, I observed, clinging to the stems of weeds, several specimens of what appeared to be the bodies of grasshoppers with the wings of the common sulphur-yellow butterfly attached to them. Such a combination aroused my curiosity, but a close examination proved them to be specimens of the common black-winged grasshopper which had just moulted for the last time and spread out their soft wings to dry. The inner wings, instead of being black, were light yellow, but in three or four hours thereafter had changed to their usual color. The cast off skins were close by and were much smaller than the insect. Like crayfish they had shed the entire outer skin in one piece, pulling out the legs

Fig. 60—Grasshopper shedding its skin.

(*a*, Young ready to change; *b* and *c*, the skin splits along the back and the adult emerges; *d*, adult drying out; *e*, perfect insect. After Riley.)

and antennæ much as a person pulls a foot out of a boot.

However, all grasshoppers do not pass the winter in the egg state. Three or four species hatch in early autumn and the young in various stages can, in suitable

Winter Grasshoppers. localities, be seen jumping vigorously about on any warm sunny day in mid-winter. If their presence at such a season comes to the attention of a newspaper reporter, the press of the entire State is apt to teem with warnings of a coming grasshopper plague, of which the youngsters are thought to be the advance guard. These hibernating young are the first to reach maturity the next spring, usually becoming full grown about the 20th of April.

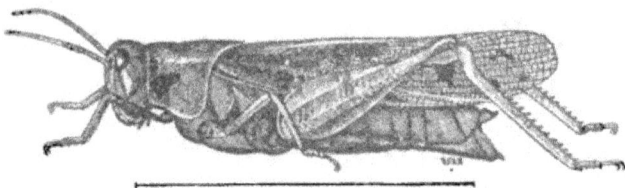

Fig. 61—Coral-winged Grasshopper, *Hippiscus tuberculatus* (Pal. de B.). (Survives the winter in young stage and reaches maturity in April. After Lugger.)

Again, nine species out of our fifty-one pass the winter as mature insects. They are our smallest grasshoppers, all being, when full grown, less than half an inch in length; gray or blackish in color; and

Grouse Grasshoppers. with the hard upper crust of the thorax extending the full length of the body and covering the wings. They are called "grouse grasshoppers," and during cold weather they hide beneath the loose bark of logs

or beneath the bottom rails of old fences.   On the first
warm days of spring they can be collected by hundreds
from any grass-covered hillside having a sunny southern exposure.

A grasshopper has five eyes, three small simple
ones, and two large compound ones.   Each of the
latter is composed of many thousand six-sided facets
or parts, in each of which a single filament of the
optic nerve ends; yet it is claimed that with all these

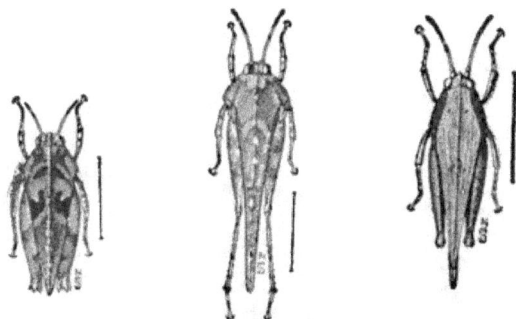

Tettix ornatus (Say).   *Paratettix cucullatus* (Burm).   *Tettigidea lateralis* (Say).

Fig. 62.—Grouse Grasshoppers.   (After Lugger.)

eyes the insect sees but poorly, being guided rather
by the sense of smell than by that of sight.   There is
no nose, the sense of smell being located in the feelers
or antennæ; while the ears, instead of being in the
head, are on the basal ring of the abdomen.   Ten
small openings on each side of the body lead into
tubes which branch and ramify through all its
portions.   Through these the air passes and comes in
contact with the blood vessels which lie alongside of
the air tubes in many parts of the body.

As is well known, the male of each species of grass-hopper has the power of making a peculiar noise, or "stridulation," as it is called. In most species it is made while on the ground and is produced by rubbing

"Songs" of Grasshoppers. the inner surface of the hind leg against the outer surface of the front wing. In those species which fly much it is made while on the wing, or just when rising from the ground, by rubbing together the upper surface of the front edge of the hind wings and the under surface of the front wings. By paying close attention the observer can soon learn to distinguish each species by its peculiar note. Only the males have musical organs, which is quite the reverse among some animals higher in the scale of life where the females make most of the music and oftentimes much of the noise. The female grasshoppers, however, make up for their lack of musical abilities by their greater bulk, as they are always much larger than their better (?) halves.

Nineteen out of our fifty-one species seldom use the wings in moving from place to place, but leap vigorously when disturbed. Among them are seven species of "short winged" grasshoppers whose wings have become rudimentary in the past through long disuse, so that in the perfect insect of to-day they are less than half the length of the abdomen.

Fifteen kinds, the most common of which is the black-winged or "Carolina grasshopper," use the wings almost wholly in their journeyings and often fly long distances when flushed. Their hind legs are used only in giving themselves an upward impetus

from the ground and hence are much smaller propor-
tionally than are those of the group of "hoppers"
which leap rather than fly, while their wings are
much longer and stronger.   To this group of "flyers"
belongs our largest and most handsome species,
the "American grasshopper," *Schistocerca americana*
(Drury).

In the season of 1893 this species was unusually
common in Vigo County, from the fact that a large
number of adults were blown in by a high wind which
prevailed on the night of April 11. No mature specimens had ever before been noted in that vicinity'

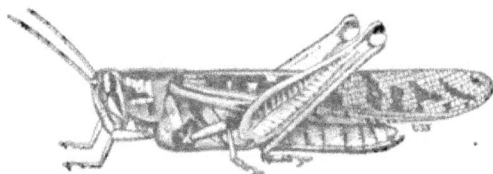

Fig. 63—American Grasshopper.
(Male. After Lugger.)

earlier than the middle of June, but on the morning
after the storm mentioned hundreds were seen on the
streets of Terre Haute.   They had come sailing in on
the wings of the wind from some distant point in the
south-west where they had passed the winter in the
mature state or as an advanced form of the young.

In Indiana there are two species which far outrank
all others in numbers and in the injury which they do
to grass and growing vegetation in general.   The
larger of these is the "lubberly grasshopper," *Melanoplus differentialis* (Uhler), a clumsy, thick-set fellow,
which is found by thousands along fence rows and

the borders of cultivated fields, especially those of the river bottoms, where they feed upon the greater ragweed or horse-weed. On October 2d, 1894, vast numbers were seen along the edge of a field of low-

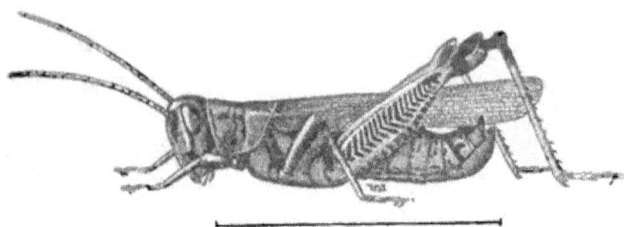

Fig. 64—Lubberly Grasshopper.
(Male. After Lugger.)

land corn, the leaves of the marginal rows of which they had almost wholly destroyed. When a stalk was approached they did not desert it but dodged quickly around to the opposite side, much as a squirrel does around the trunk of a tree when pursued. If, however, one took alarm and jumped, all the others in the immediate vicinity did likewise.

The most common and most injurious species found in the State is much smaller and is known as the "red-legged grasshopper," *Melanoplus femur-rubrum* (DeGeer), on account of its hind shanks or tibiæ being blood-red. It often destroys the second crop of clover in many parts of the State, besides doing much injury to young corn and other growing crops.

The Kansas or Rocky Mountain grasshopper does not occur in Indiana. Contrary to the general belief

Fig. 65—Red-legged Grasshopper.
(Male.)

it is not a large, robust species, being but about the

**The Kansas
Grasshopper.**
same size as our red-legged grasshopper, and bearing to the latter a close general resemblance; so close in fact that only specialists can readily tell them apart.

Millions of dollars of damage was done in the Western States by these small insects in the summers of 1873 and 1875. Migrating in vast clouds from one part of the country to another they would fall upon a

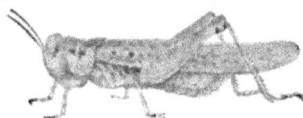

Fig. 66—Kansas Grasshopper.
Melanoplus spretus (Uhler).
(Male.)

cornfield and convert, in a few hours, the green and promising acres into a desolate stretch of bare, spindling stalks and stubs. In the words of the prophet Joel: "The land was as the garden of Eden before them, and behind them a desolate wilderness; yea, and nothing did escape them."

The sound caused by their approach or flight was almost deafening, and has been graphically described as follows:

> "Onward they came, a dark continuous cloud
>   Of congregated myriads numberless,
> The rustling of whose wings was as the sound
>   Of a broad river headlong in its course
> Plunged from a mountain summit, or the roar
>   Of a wild ocean in the autumn storm,
> Shattering its billows on a shore of rock."

Each kind of grasshopper has its favorite resort where it is to be found in greatest abundance, although some of them are common enough anywhere. For example, there are three or four species of pea-green grasshoppers which are found only among the dense green grasses and sedges along the margins of ponds and lowland streams. There, as long as motionless, they are invisible, and there they flourish in peace and countless numbers.

The Kentucky blue-grass and the different kinds of meadow grasses are a darker green, and, where rank, turn brown early in the autumn. The different species of "short-winged" grasshoppers, and many others whose hues are olive green or brown, find in the fallen clumps of these grasses places of hiding well suiting their color as well as an abundance of food well suiting their taste.

At the Goose Pond, nine miles below Terre Haute, occurs a species of grasshopper, *Leptysma marginicollis* (Serv.), which has never been recorded elsewhere north of Florida. Its occurrence in Indiana can only be accounted for by the presence of the broad and sheltering valley of the Wabash within the confines of which it finds a climate and a vegetation congenial to its wants. If its habits be the same elsewhere as in Indiana, the name "grasshopper" is for it a misnomer, for here it is never seen on the grass or ground, and never hops when disturbed, but moves with a quick and noiseless flight for twenty or more feet to a cylindrical stem of sedge or rush on which

Fig. 67—A Pea-green Grasshopper. *Dicromorpha viridis* (Scudder). (After Lugger.)

16

it alights. The instant it grasps the stem it dodges quickly around to the side opposite the intruder. Then, holding the stem firmly with its short front and middle legs, it draws its slender hind legs close up against the body and hugging its support as closely as possible, remains perfectly motionless. Its body is almost cylindrical, and being of the same general color as the stalk of the plant on which it rests, it is almost impossible to detect it, unless one sees ex-

Fig. 68—Florida Grasshopper.

actly where it alights. Eight times out of ten a person, by approaching quietly, can reach his hand about the plant stem and grasp the insect. Its habits excellently illustrate the so-called " protective mimicry" of form and coloring, as it always seems to choose a cylindrical object, and one similar to its own color, before alighting.

Let us now briefly notice the habits of our ash-brown, black-winged friend, the " Carolina grasshopper." To the casual observer he appears to be our most common species but there are a dozen which are more abundant. His numbers appear multiplied because he frequents the highways and byways of man rather than the pastures and meadows where other grasshoppers are wont to congregate. Moreover, when disturbed, he more often betakes himself to the bare earth than to the green grass. " Why this absurd taste?"—asks the person uninitiated in the doings of nature's objects. For the simple reason that the dust of the roadside and the gravel ballast of the railway correspond so closely with the color of his back that his best friends and worst enemies will overlook him if he will only remain quiet. Yea, even that sharp-eyed connoisseur of grasshopper tidbits, the turkey gobbler, oftentimes walks right over him mistaking him for a wayside pebble.

Thus, by choosing their resorts to suit their colors, or rather, in the course of ages, changing their colors to suit their environments, grasshoppers have been enabled each

Fig. (?)—Carolina Grasshopper.
Dissosteira carolina (L.).
(After Lugger.)

year to wax, grow fat, and replenish the earth with their progeny, at the same time deplenishing it by reason of their enormous numbers and their insatiable appetites.

In the latitude of central Indiana, the heyday of the mature grasshopper's life lies between the dates of mid-August and mid-October. Then it is that their love calls are the most numerous; that their ambitions in life are satisfied, and that the eggs, destined to carry the race through the rigors of an approaching winter, are deposited in the earth. As the cooler days of November draw nigh their ranks grow rapidly less, and yet, with the exception of one place, their dead bodies are seldom met with. That exception is the top of tall weeds, where, oftentimes even so early as September, the bodies of the more common species are seen, their limbs tightly clasping the branches or leaves of the plant on which they rest. Why this position is taken by them before death I can not say. Of course it is nonsense, but I have often asked myself the question: Is it assuming too much of them to suppose that, having lived their allotted time, or being preyed upon by some invisible but insidious animal or plant parasite, and feeling their death throes coming on, they choose to fly or climb to the most elevated position available, there to take a farewell view of their summer's home?

Fig. 70—Two-striped Grasshopper. *Melanoplus bivittatus* (Say). (Killed by a fungus. After Lugger.)

WEEDS IN GENERAL AND OUR WORST WEEDS IN PARTICULAR.

What is a weed? How does it differ from a wild flower? These questions are often asked by persons who are beginning the study of botany; and pupils have been known to put aside a specimen with a look of disgust, saying at the same time that they "did not want to study that nasty weed."

There is, of course, no difference, botanically speaking, between a weed and a wild flower, save that of comparative abundance. Some of our most common weeds are among the most handsome of our wild flowers; for example, the iron-weed, thistle and ox-eye daisy. They well illustrate the truth of that old saying that "familiarity breeds contempt," for we have become so familiar with their appearance that we daily pass them by unnoticed. Were they as rare as the showy orchis and wild columbine they would no longer be called "weeds," but "wild flowers," and would, perhaps, be cultivated for ornament; just as among half the collections of house plants in Indiana are found species of cacti which are by no means rarities to the natives of Texas and New Mexico.

A weed has been defined as "merely a plant in the wrong place," but Grant Allen, a noted English botanist, in speaking of this definition says that it is far

more than that, and that the term "weed" implies
something further than mere abstract hostility to the
agricultural interest; it "implies a certain ingrained
coarseness, scrubbiness, squalor and sordidness, besides
connoting, in nine times out of ten, some stringiness
of fibre, hairiness of surface, or prickly defensive
character as well." Of most weeds this is true, but it
is the possession of just such characters that has
enabled them to succeed so well in crowding out and
displacing other wild and cultivated plants which,
perhaps, were in the right place, and in taking, if
unmolested, entire possession of the soil.

Weeds are, of course, among the worst enemies
with which the farmers and market gardeners have
to deal, and the questions of eradication, quarantine,
and the like are becoming each year of greater state
and national importance.

Each locality, and each farm for that matter, has
its worst weed, the species depending upon climate,
soil, method of cultivation, etc. Indeed there is no
spot on earth, unless it be a desert, where they do not
abound. A few years ago Mr. Byron D. Halstead
asked, through the Botanical Gazette and other peri-
odicals, that a list of the 20 worst weeds, in the order
of their injuriousness, in any locality or territory in
the United States, be sent to him. He tabulated the
reports which he received from all parts of the country and found that 34 species had been mentioned at
least five times in the lists. Of these 34, no less than
32 are known to occur in Indiana, though some of
them only in a few localities of the State. Taking
from the list of 34 the 20 species which were consid-

ered the most injurious, all of them are represented in Indiana; although the order of injuriousness here, is, in my opinion, not the same as that given in Mr. Halstead's list.

The Canada thistle and couch or quick grass, which are ranked first and second in his list, are, as yet, found only in restricted areas in northern Indiana. The common names of the remaining 18 species of the list in the order given are as follows : Cocklebur, bur-grass, crab or finger grass, rag-weed, field sorrel, pig-weed, horse-weed or great rag-weed, fox-tail, lamb's-quarters, ox-eye daisy, purslane, curled dock, barn-yard grass, bind-weed, shepherd's purse, common thistle, burdock and jimson.

Each of these weeds has its favorite locality of growth; for instance, the great rag-weed, cocklebur and bind-weed are found almost exclusively in the rich alluvial soil of the bottoms; burdock and jimson-weed in waste grounds about stables and old dwellings; rag-weed and foxtail in cultivated upland soil; bur-grass along the sandy banks of streams, and so on through the list.

To the botanist, one of the most interesting points in connection with these 20 weeds is the fact that 15 of them are introduced or foreign plants which have become naturalized from Europe or Asia; the two rag-weeds, bind-weed, bur-grass and couch-grass comprising the five American or native species. Indeed, America seems to be not only the "home for the oppressed of all nations," but her soil seems to suit exactly those weeds which are the off-scourings and refuse of civilization in all countries. Grant Allen

expresses it well when he says that in " civilized, cultivated, and inhabited New England, and as far inland
at least as the Mississippi, the prevailing vegetation
is the vegetation of central Europe, and that at its
weediest. The daisy, the primrose, the cowslip and
the daffodil have stayed at home; the weeds have
gone to colonize the New World. For thistles and
burdock, dog-fennel and dead-nettle, hound's-tongue
and stick-seed, catnip and dandelion, ox-eye daisy
and cocklebur, America easily licks all creation. All
the dusty, noisome, and malodorous pests of all the
world seem there to revel in one grand, congenial,
democratic orgy."

Of the plants described in Gray's " Manual of Botany," as growing east of the Mississippi and north of
North Carolina and Tennessee, 293 are introduced
species, 27 of which are natives of tropical America,
the remaining 266 having found their way here from
Europe; while 342 other species are common to the
north-eastern United States and Europe. Thus an
American botanist crossing the Atlantic could find,
growing indigenously in Europe, no less than 608
species of plants which he was accustomed to see at
home and they among the most common ones found
here.

An interesting history of the numerous ways in
which the Old World weeds have been introduced into
this country could, no doubt, be written if one had all
the facts. One instance of how a single species found
its way from Germany to this State will serve as a
type of the method of introduction. While a student
at Indiana University, I was engaged for a time in

working up the flora of Monroe County, and one day happened upon a strange species of the *Compositæ*, or sunflower family of plants, growing in the new college campus. It proved a puzzler, and after spending the better part of my spare time for two days in endeavoring to find its name, a specimen was sent to Professor Dudley, the botanist at Cornell University, for identification. He, having traveled in Europe, immediately recognized it as a pernicious weed common on the continent, but not before reported from any part of the United States. The next question was, how had it found its way into that remote corner of Indiana? It was easily answered. A new supply of glassware for the chemical department of the University had, the fall before, been purchased in Germany, and the straw in which it had been packed was thrown on the ground and left for a day or two on the very spot where the plant had afterwards appeared. The three specimens which sprang up were destroyed before maturing their seeds and the spread of the weed throughout the country was thereby prevented.

But all the weeds introduced into this State in recent years are not foreigners or descendants of foreigners. The supply of new species from Europe is about exhausted, and the great plains of the west and southwest, admirably adapted by nature for the evolution of weeds of cultivation, are rapidly sending eastward their own rich contingent to compete with the trans-Atlantic types for the mastery of our soil. Twenty years ago there started eastward from the base of the Rocky Mountains, the bristly cone-flower and the fetid marigold, two members of the great

Compositæ order. To-day they have a foothold east of the Mississippi, along every roadside and in every meadow suitable to their growth. Two specimens of Texan nettle (*Solanum rostratum* Dunal) were recently found in a field north of the city of Terre Haute. This plant is given in Prof. Halstead's list as one of the 34 worst weeds in the United States. Its habitat, or usual range, is given in the botanies as "Plains of Nebraska to Texas," but it is rapidly moving northward and eastward, and unless checked in its course will soon bring dismay to thousands of farmers who know nothing of its pernicious habits.

These plants in their eastward migration are well up to the spirit of the times. They travel by railway. The seeds are carried either in the coats of cattle or sheep or in the food which supports them on their journey. Our great railways run east and west and the bared soil alongside the tracks furnishes excellent seed beds, where, if dropped, the seed may sprout and the plant grow unmolested, until it gets a chance to take another step in advance. The botanist has learned their ways of migration and knows that if he wishes to find strange species his best pathway will be alongside the railways.

Taking the term "weed" in the sense of useless plants growing wild in cultivated grounds, pastures and meadows, of sufficient size to be easily noticeable, and of sufficient abundance to be injurious to the farmer, 91 species were found by myself in Vigo County and as many may be found in almost any county in the State. Of these, according to Gray, 33 are of European origin; eight are from tropical

America; two from Asia; the remaining 48, natives of the United States.

In my opinion the following list comprises the 20 worst species of weeds growing in Indiana, and named in the order of their injuriousness. Both common and scientific names are given, together with the original home of each species:

1. Rag-weed, *Ambrosia artemisæfolia* L.—United States.
2. Foxtail, *Setaria glauca* Beauv.—Europe.
3. Iron-weed, *Vernonia fasciculata* Michx.—United States.
4. Great Rag-weed, *Ambrosia trifida* L.—United States.
5. Pig-weed, *Amarantus retroflexus* L.—Tropical America.
6. Horse-weed, *Erigeron canadense* L.—United States.
7. Cocklebur, *Xanthium canadense* Mill.—United States.
8. White-top, *Erigeron annuus* L.—United States.
9. Lamb's-quarters, *Chenopodium album* L.—Europe.
10. Common Thistle, *Cnicus lanceolatus* Hoffm.—Europe.
11. Field Sorrel, *Rumex acetosella* L.—Europe.
12. Purslane, *Portulaca oleracea* L.—Europe.
13. Bur-grass, *Cenchrus tribuloides* L.—United States.
14. Beggar's Ticks, *Bidens frondosa* L.—United States.
15. Prickly Lettuce, *Lactuca scariola* L.—Europe.
16. Crab-grass, *Panicum sanguinale* L.—Europe.
17. Jimson-weed, *Datura stramonium* L.—Asia.
18. Smart-weed, *Polygonum persicaria* L.—Europe.
19. Bracted Bindweed, *Convolvulus sepium* L.—United States.
20. Corn Cockle, *Lychnis githago* L.—Europe.

Of these it will be seen that nine are native; nine are from Europe; one from tropical America, and one from Asia. The first, third and fourth are native, so that all in all the American weeds have held their own quite successfully in the Hoosier State.

And now, we suppose, the question naturally arises: "What is the best method of ridding ourselves of these weeds now that they are here?" In my opinion

there is no method. They are here and here to stay.
The farmers of the future must wage an eternal warfare against them, for they have secured a foothold
which can not be entirely overcome. True, a new
species possessing advantages which will enable it to
crowd them out, may, in time, appear, but such a
change would very likely be for the worst.

There was a time when but one, two, or a dozen
plants of each of the foreign weeds existed in the
State. Then was the time to have successfully quar-
antined that species by destroying those pioneers.
The few persons on whose lands they appeared neg-
lected them, and every gardener, every farmer, yea,
every land owner in the State, must henceforth, now
and forever, pay the penalty of that neglect by continued hoeing, plowing and mowing to keep these
alien weeds in subjugation.

As long as the rudiments of botany are not taught
in the common schools the average farmer will be
unable to tell whether a new plant which has made
its appearance upon his land should be allowed to
grow or not; in fact, in many instances, he will not
know that a new plant is there until it becomes too
abundant to be easily overcome. Put a high school
into each township in the State; teach the elements
of botany therein and then, and not till then, may we
hope that the farmers of the future will be on the
lookout for all new plants; will be able at once to
judge their relative injuriousness; and will destroy,
before they have time to ripen their seeds, those
species which, if allowed to spread, would become a
curse to the State.

TWELVE WINTER BIRDS.*

I.

The woods and fields in winter are not the silent, deserted places which most people believe them to be. Any person who will look with both eyes and listen with both ears can detect in them many forms of life —occupying many strange and wonderful positions— and gaining a livelihood in many quaint and curious ways.

With the wind blowing at the rate of thirty miles an hour, the air thick with falling flakes of snow, and the temperature 15 degrees Fahr., or less, one is, perhaps naturally, disposed to stay indoors and take it for granted that all the birds have long since departed for the sunny south. But herein mankind sadly errs, for even during such days, both in the woods and fields, there are birds and birds. The ornithologist, strolling for a mile or two beyond the city limits can on such a day, devoted solely to the observation of his feathered friends, usually detect 30 or more species, while fully 28 additional kinds are found in the State during the winter season. These may be classified among three groups:

First.—PERMANENT RESIDENTS, or those which rear their young here; they or other individuals of their

*First published in Terre Haute Gazette; December, 1893–March, 1894.

species remaining with us throughout the year—
the quail, crow and jay-bird being familiar examples.
Of the 321 birds known to occur in Indiana, 33 belong
to this class.

Second.—WINTER RESIDENTS. These nest in the
northern regions and come down each season in late
autumn to spend the winter months with us and cheer
us up with their merry chirps, but disappear northward again at the approach of spring. The slate
colored snow-bird and the tree sparrow are the most
familiar and abundant of the eleven species of this
group which occur in Indiana.

Third.—WINTER VISITORS. These are birds either
from the north or north-west which often drop in
upon us to spend a week or two when King Boreas
reigns supreme, and the mercury marks daily the zero
point or below. The great snowy owl, the red cross-
bill and the golden eagle are examples of the 14
species of this class which have been noted in the
State.

In the wooded portions of Indiana the woodpeckers
are among the most noticeable and interesting members of our winter bird fauna. Few are the days
from November to April when their peculiar calls

**The Wood-
pecker Family.**
and rapid tapping in search of food
may not be heard. About 250 kinds
of woodpeckers are known, only eleven
of which are found in the United States east of the
Mississippi river. Of these, seven occur in Indiana,
five of the seven being permanent residents; one, a
migrant, is seen here only in spring and fall; while the
other one is a summer resident, but often remains in

small numbers throughout the winter, especially in the southern half of the State.

All belong to the family *Picidæ*, a word derived from the Latin *picus*, a "woodpecker." The characters by which each member of this family may be easily known are the stout, straight bill, fitted for hammering or boring into wood; the long, barbed tongue which, like that of the snake or toad, can be darted from the mouth for the purpose of catching insects; the toes in pairs, two in front and two behind, and armed with strong, compressed claws, thus enabling

Fig. 71—Zygodactyle or "yoke-toed" foot of woodpecker.

the bird to get a firm hold upon the trunk or limb of tree; while the tail feathers are not soft and rounded like those of other birds, but are very stiff and pointed at the end, thus enabling the owner to use the tail as a brace and so keep from toppling over backwards while delivering its rapid and powerful blows. The flight, too, of these birds is peculiar, being a sort of wave-like or undulatory progression, instead of a movement directly forward on one level.

One of the most interesting of the five species of woodpeckers found in the State throughout the year is the subject of the present sketch. It has a number

of common names, the most familiar of which is the
"red-bellied woodpecker."   Others are "zebra bird,"

The Red-bellied Woodpecker. "woodchuck," or "chuck," "Carolina woodpecker" and "checkered woodpecker;" while its scientific name is
Melanerpes carolinus (L.).   The first part of this name,
Melanerpes, is the name of the genus and corresponds
to our surnames of Jones, Brown, etc.   It is derived
from two Greek words and means "black creeper."
The second part, *carolinus*, corresponds to our given
names of John, Mary, etc., although it is always written after the sur or generic name.   It is derived from
"Carolina," from which region the first specimens were
described by Linnæus, the noted Swedish naturalist
of the last century.

Thus, each kind of bird, as well as each kind of
organic object which has been noted and studied by
man, has a double Latin name by which it is known
to naturalists in every country on earth; while it may
have a dozen common names in the restricted locality
where it is found.   There are thousands of birds of
this name in Indiana, but they are all very nearly
alike—that is of one species—and hence have the
same Latin name; while those woodpeckers which
differ materially in size, color, etc., have different Latin
names, as we shall see on the following pages.

The red-bellied woodpecker is about ten inches in
length from tip of bill to end of tail.   The back and
wings are very prettily barred with narrow black and
white bands, whence the less used but more suitable
name of "zebra bird."   In the male the entire upper
surface of the head and neck is a bright scarlet-red,

while in the female the crown is ash colored, with the
forehead and nape of the neck scarlet. The whole
under part of the body is a grayish ash with a tinge
of red on the belly.

The zebra bird is the hermit among our woodpeckers. He scorns the companionship of other members

Fig. 72 Red-bellied Woodpecker.

of his family and delights to be alone or in the company of his mate only. During the winter and early
spring he may be found along the borders of woodlands, where in the tops of the tallest trees he industriously seeks his food. At intervals he pauses in his
work, raises his head and looking around, utters a

17

loud "char-char" or "chuck, chuck, chuck," and then immediately resumes his pecking with increased vigor. Rarely in winter, when insect food is scarce, does he visit the farmer's corn crib and, gaining ingress between the logs or rails, helps himself to a bountiful supply of food.

As the nesting season approaches he, with his chosen mate, seeks the privacy of the deepest woods, where high above the ground in some decaying limb or trunk of tree a place for a nest is excavated. The eggs are four to six in number, pure white in color, and about 1x.87 inches in size.

To the farmer who owns timber land this woodpecker is one of the most beneficial of birds. The number of kinds of insects which prey upon our native forest trees is surprisingly large. The oak harbors between 500 and 600 species; the hickory, 140; the walnut, 70; the elm, 25 to 30, and the maple at least 15. To the presence of these insects, more than to anything else, is the stunted growth and early decay of our timber due. Thousands of wood-boring beetles, caterpillars, plant lice and young saw-flies are annually destroyed by the zebra bird. On this account the farmers who have not already made his acquaintance, should at least seek to know him by sight, and always protect him and his young from their now most dangerous enemies, the small boy with his musket and the city sportsman with his shotgun.

In autumn, especially, the woods and fields throughout the State are full of these self-styled "hunters," ready at a moment's notice to bring down any bird larger than a sparrow, any animal above a field mouse

or a chipmunk. No feeling of sorrow ever enters
their minds as they gaze into the eyes of some cruelly
wounded bird or animal and see the life force, which
they can never recall, slowly ebbing away. The
woodpeckers, in particular, suffer from their heedless
shooting. Hundreds of them are daily shot down
and left where they fall, for the sole purpose of showing the hunter's skill in marksmanship, or to satisfy
that craving desire to kill objects below him in the
scale of life which blinds him to every feeling of pity,
every sense of remorse.

To those who have been accustomed to pass the
cold season in the city, exempt for months from the
pure, bracing, country air, I would say : Go forth and
study nature on some sunny day in mid-winter. You
will not find the woods full of thrushes, warblers and
other songsters ready on every hand to greet you.
They are effeminate birds, joyous only in the presence
of plenty of blue sky and sunshine, and Jack Frost
has long since driven them to seek a sunny, southern
clime where food is more plentiful than here. But
you will find their rugged cousin, the zebra bird,
clinging to the side of a dead stub and deriving pure
inspiration therefrom. You will see him fluff his
feathers about his bare toes to keep them warm while
he makes the woods reverberate with his cheery
call—and from him, if you are wise, you may learn a
lesson of happiness and contentment.

II.

He, who would become acquainted with our common birds, must seek them in their chosen haunts; for in the ages of the past each kind has become fitted or adapted to live a certain life—to seek its food in a certain place. In the spring, keen-eyed, sharp-eared robins hop leisurely over the surface of our lawns and gardens and drag earth-worms and cut-worms galore

Fig. 73 - Robin. (After Beal.)

from their hiding-places; long-legged snipe wade the shallow water along the margins of ponds and streams and probe with their long, soft bills the muck and mud in search of worms and leeches; web-footed wild ducks swim and dive in ponds and rivers, silting through their broad lamellate bills the mud and ooze for small fish, shrimps, and other water-loving forms;

wide-mouthed swallows cleave the air in varied circles, snapping up on the wing the myriads of unlucky
insects which they meet; but most wonderfully, most
strikingly adapted for the life they lead are the members of the woodpecker tribe. One seeks them not
on the smooth lawn, nor along the margin of pond
and stream, nor in the blue vault of heaven above:
for, as their name indicates, they are peckers of wood,
and on trees, fences or wooden objects only, are they
found. Darwin, in his epoch making book, "The
Origin of Species," calls attention, time and again,
to their marvelous adaptation of beak and tongue, of
toes and tail. Couple with these a keen eye, a sharp
ear and a long, strong wing, and we have the hardy
denizen of the woods, who braves unflinchingly every
clime and finds a plentiful living where other birds
would quickly starve.

Of the five species of woodpeckers which spend the
winter months with us there are two which are often
confounded by the person who is beginning the study
of birds. At a distance they appear to be colored
exactly alike, both being black above with a white
stripe running lengthwise down the back and numerous white spots arranged in crossbars on the quills of
the wings. There is a narrow white stripe above the
eye, and below it a broader and longer one extending
beyond the eye, almost to the white stripe on the
back. In the full grown male there is a crescent
shaped scarlet band across the back of the head which
is always lacking in the female. The under surface
of the body is a uniform grayish-white.

In size alone are the two species readily separated;

the larger or "hairy" being more than two inches
longer than the "downy;" the average length of the
former being nine inches from tip of bill to end of
tail, while that of the latter is but six and a half
inches. Only when we have them in hand do we detect any other difference than that of size. Then, by
spreading out the tail, we find its outer feathers to be
pure white in the "hairy," while in the "downy" they
are cross-barred with black near the tip.

These birds were both first described by Linnæus,
who gave to the larger one the name of *Picus villosus*
and to the smaller one that of *Picus pubescens*. The
word "*Picus*" means "woodpecker," while *villosus* and
pubescens both mean "downy" or "covered with very
fine soft hairs." Later writers have divided up the
genus *Picus* and have assigned certain of its members
to other genera. Among these are the two species
now in hand which, with a few others not occurring
in Indiana, are placed in the genus *Dryobates*, which
means "oak walker." The common names "hairy"
and "downy," have evidently been derived from the
specific names, and signify no difference whatever in
the character of the plumage of the two birds.

The hairy woodpecker, *Dryobates villosus* (L.), has
many local names in different parts of the country,
The Hairy Woodpecker. chief among which are the "Virginia
woodpecker," "big sapsucker," "big
spotted woodpecker," and "big guinea
woodpecker," the adjective "big" being prefixed to
distinguish him from his smaller cousin which in color
he so closely resembles.

He is less sociable than the "downy," frequenting

more often than that species the depths of the woods;
though hunger often drives him in mid-winter to the
outskirts of the city, where, in the tops of the many
decaying shade trees, he finds the struggle for existence
between himself and his kin less deadly than in his
native wood. A careful examination of 82 stomachs

Fig. 74—Hairy Woodpecker. (After Beal.)

of this species made by the Bureau of Ornithology at
Washington, showed that 68 per cent. of the food was
animal matter, largely the remains of injurious insects;
while 31 per cent. was of vegetable origin, consisting
mainly of wild cherries, grapes and the berries of the
dogwood and Virginia creeper.

On one bright December morning, I watched, for a half hour or longer, a hairy woodpecker which had alighted near the middle of an upright limb of an oak. Instead of moving upwards, as is his usual wont, he hopped backwards an inch or two at a time for twenty or more feet, peering intently, on both sides as he descended, into every crevice and cranny in the bark. Not a sound did he make meanwhile, though a saucy tom-tit was uttering its "*dee-dee-dee*" within a few feet of his head and a zebra bird on an upper limb gave forth at intervals its loud alarm note of "*char-char.*"

Reaching the base of the limb the "hairy" flew a foot or two to another and began to hop up this at the same rate of speed as he had descended the first. At short intervals he sounded the wood with his bill, listening intently the while, and at last, I suppose, the percussion was satisfactory; for, fixing more firmly his stiff tail feathers against the tree as a prop, he rose to the full length of his short powerful legs, and drawing back his body, head and neck to the farthest extent, he dashed his wedge-shaped bill home with all the force of his entire bodily weight and muscle. How the bits of lichen, bark and fragments of half rotten wood came tumbling down and how handsome he looked with the scarlet cap on the back of his head, and the white central stripe contrasting so vividly with the glossy black of his back!

At last the reward came in the shape of a good fat grub which was quickly drawn from its hiding place by the long barbed tongue of the bird. After swallowing this prize he was evidently well pleased with

its taste for he uttered for the first time his usual call,
"*plick-plick*," and then began once more his active
search, pecking, pecking, hammering on his upward
course.

In this way his daily life is spent from November
to March, often enduring ice, frost, snow, sleet, rain
and hail, but cheerfully meeting them all undaunted.
When the warm April sunshine enlivens the woods,
and insect food can be had for the asking, he seeks
for himself a mate, and together they excavate a nest
in the terminal limb of a lofty beech or in the top of
some tall oak snag. Here the eggs, five or six in
number, are deposited. They are pure crystal white
with an ivory texture and are .85 x .65 of an inch in
size.

III.

The "downy woodpecker," *Dryobates pubescens* (L.),
appears to be much more plentiful than the "hairy,"
but this is doubtless due to his sociable qualities as
he seems to seek, rather than shun, the society of

The Downy Woodpecker. man. One does not so commonly find
him in deep woodlands as in the
orchards and the borders of woods
near farm houses. Often, especially in the spring-
time, they are seen along roadsides, clinging to the
upright stakes of the old worm fences, watching the
plodding ploughman at his toil and scolding mildly
if he comes too near. When the blasts blow cold,
and the soggy limb of the forest tree is frozen hard,
one may sometimes see a downy on the dead limb of a

shade tree in the very heart of the city, where, though surrounded by a bedlam of noises, he pecks away as industriously and contentedly as he does when in the depths of the forest. But they frequent the city much more seldom now than when that despised alien, the ubiquitous and aggressive English sparrow, was a stranger to our land.

The downy is known by several other common names; chief among which are the "little spotted woodpecker," and the "little sapsucker." The latter name is, however, a misnomer, as neither he nor the hairy ever bore into trees for the purpose of gathering sap as does their cousin, the yellow-bellied woodpecker, *Sphyrapicus varius* (L.). The last named bird is a migrant, seen here only in spring and fall, and the holes he bores are small, shallow and close together, in regular circles about the tree; while those made by the downy and hairy in search of insect food are scattered irregularly over the surface of the limb or trunk.

Fig. 75—Head of Downy Woodpecker.

On a recent October morn I watched a "downy" carving with his strong, chisel-shaped bill a cavity in a dead snag, probably one which he intended using as a winter shelter. Clinging to the edge of the hole, he would reach in until only his tail was visible, give three or four vigorous pecks, then draw forth his body and head and "spit out," or rather cast aside the frag-

ments of decayed wood he had broken loose. Glancing hurriedly about on all sides, he would again dodge
in for a new series of blows. His feet remained in
one place, but when at work his body was all motion;
his tail, as he pecked, bobbing up and down against
the wood, two or three inches below the opening of
the hole. Thus he worked for twenty-eight minutes,
when a slight movement of mine frightened him and
he was up and away. The bottom of the cavity was
five inches below the opening. Small feathers were
found in it on several occasions during the following
winter, showing that it was used as a resting place,
probably at night.

To the farmer and fruit raiser the downy woodpecker is very valuable as an enemy of caterpillars.
It preys upon these insects in all stages. It drags the
butterflies and moths from their hiding places; licks
up by the score their deposited eggs; feasts for days
in early summer upon the caterpillars themselves, and
in autumn and winter bores neat round holes into the
sides of the leathery cocoons and extracts the contents
thereof. It is only when insect food becomes very
scarce that it deigns to feed upon such dry forms of
nutriment as grain, seeds of grasses and the softer
nuts. From an examination of 140 of the stomachs
of this bird it was found that 74 per cent. of the food
was of animal origin.

The "downy" is sociable, not only with mankind,
but also with his feathered kin in general. Especially
in winter, does one often find him the leader or guide
of a little coterie of chickadees, kinglets, nuthatches,
and titmice—ramblers all, who go roaming in com-

pany the livelong day. Their whistles and chirps,
twitters and trills, uttered in unison, make a jubilee
in January equal to any heard in June when the birds
are supposed to be most numerous. I have listened
manifold times to such a chorus and so have many
other persons, particularly Dr. C. C. Abbott, who
described as follows one he heard on a mid-winter
morn with the mercury at zero: "The clear call of
the crested tit opened the concert. The abundant
chickadees twittered; kinglets trilled a merry rounde-
lay; snow birds chirped; a cardinal performed an inim-
itable solo; and to all the downy woodpecker was
alike attentive and drummed a tuneful accompani-
ment on the most resonant tree in all the woods."

IV.

In books we find portrayed the thoughts of man—
in nature, those of God. Tired of poring over the
former I sought the woods and fields on the afternoon
of a recent dim December day to study the latter,
and, if possible, solve some problem, ever present,
ever interesting, concerning the life history of beast
or bird, of bug or bramble.

The blue sky of the forenoon had given way to a
sombre gray which far and wide came down to meet
the earth. No insect life was abroad on such a day.
But if the insects were lacking, what of the birds?
Had the leaden sky and north wind caused them, too,
to seek a place of shelter and remain silent and invisible? As I asked the question the *err-rrh-rrh* of a
woodpecker in a neighboring beech denied the allega-

tion; while the distant but rapidly nearing *caw-caw* of a flying crow lent strength to the denial. Woodpeckers and crows! No winter day too dull and dreary—no sky too leaden and cheerless—no north wind too harsh and biting, for them to be on the lookout for food!

To my surprise and delight I found the bird in the beech to be that handsome, tri-colored fellow known as the "red-headed woodpecker." This species which, it is said, first excited the latent enthusiasm of that eminent naturalist, Alexander Wilson, and determined him to be an ornithologist, ranges over the whole of the eastern United States from the Atlantic to the Rocky Mountains. Its long scientific name is *Melanerpes erythrocephalus* (L.); melanerpes meaning "black creeper," and erythrocephalus, "red-headed."

The Red-headed Woodpecker.

The "red-head" is so different in color from all other woodpeckers that that one character alone will be sufficient to describe him for easy recognition. In adults, or full grown specimens of both sexes, the whole head and neck are a brilliant crimson; the under parts of the body, the shorter wing feathers or secondaries, and the rump are white; while the remainder of the body is a glossy blue-black. In the young the head and neck are grayish, or slightly tinged with red. The length from tip of bill, along the back, to end of longest tail feather is 8½ to 9½ inches.

The nest of the red-head is usually placed in a cavity which it excavates in a limb or trunk of tree at some distance from the ground; but often, for want of a better locality, it chips a hole in a fence stake, telegraph pole, church steeple, or cornice about the roofs of dwellings. The eggs are four to six in number,

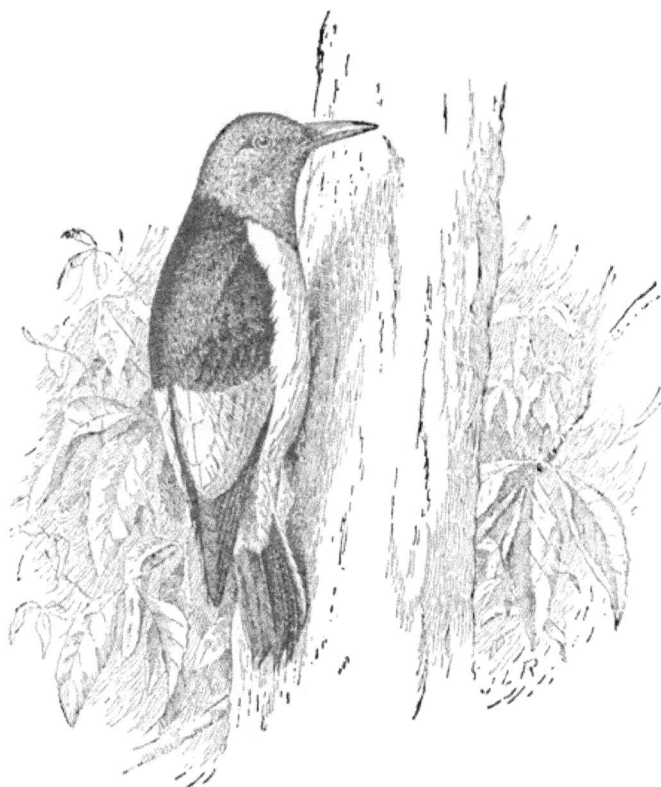

Fig. 76—Red-headed Woodpecker. (After Beal.)

elliptical in form, pure translucent white, and measure 1 x .8 inches.

In northern and central Indiana the red-head is usually a summer resident, arriving from the south

about the middle of April and remaining until mid-
October. During that time he is the most common
species of the family, holding his own with perfect
ease and ceaseless familiarity. In traveling through
the country, during the spring and summer, one hears
them screaming in the adjoining woods, rattling on
the dead limbs of trees, or on the fences, along which
they are perpetually seen flitting from stake to stake
on the roadside ahead.

In southern Indiana, and especially within the sheltered confines of the Wabash valley as far north as
Terre Haute, the red-head usually remains throughout the winter in small numbers. But in those seasons when beechnuts are plentiful he hoards or stores
up a supply of them for winter food and then remains
in large numbers, even to the northern boundaries of
the State. As soon as the nuts begin to ripen in
autumn, the birds appear to be almost constantly on
the wing, to and fro, from the beech trees to some
place of deposit. They hide away the nuts in almost
every conceivable situation. Many are placed in cavities in partially decayed trees; and the felling of any
old tree near a beech grove is certain to disclose a
pint or more of their chosen winter food. Another
favorite storing place is beneath the loose bark of the
hickory, or behind some long sliver of fence post or
rail. Sometimes the nuts are driven into the cracks
at the end of railroad ties. Again they may be found
in the crevices between the shingles on the roofs of
barns or even houses. A friend, who is a close
observer, also relates that the holes which are bored
into the beams and logs of old cribs and sheds for the

insertion of wooden pins, are often utilized as store-
houses by the birds, being filled with beechnuts as far
as the red-head can reach. They oftentimes stuff the
opening full of moss and cedar bark to hide the contents. Whether the bird remembers where each nut
is placed, or whether it trusts to luck in finding it
again is, as yet, an open question. That the red-head
lives to a large extent in winter upon this stored food
has been proven by the examination of the stomachs
of a number of specimens at that season.

The belief, so common among many persons, that
these birds are weather prophets, and, foreseeing a
long hard winter, depart early in autumn for a warmer
clime, is, in my opinion, unfounded. Irregularity of
migration, residence, and so forth, among birds, is, I
believe, due almost wholly to irregularity of a suitable food supply.

But our red-head, as he has advanced in civilization,
has developed a taste for other-articles of food besides
nuts and insects, and has become, in the words of
Maurice Thompson, "a cider taster, a judge of good
fruits, a connoisseur of corn, wheat and melons, and
an expert fly catcher as well." He is excessively fond
of ripe mulberries and wild cherries, and in their
season spends much time in winging his way to and
from these trees. I have often seen him dart from
the top of a mulberry or other tree, and with unerring flight catch upon the wing some unlucky insect,
and then immediately return to his diet of fruit once
more.

During the summer months he chooses the finest
cherries, grapes and apples for his dessert and thus

oftentimes brings upon himself the unqualified rage
of the farmer or fruit grower. When Indian corn is
in its rich, succulent, milky state he attacks it with
great eagerness, opening a passage through the numerous folds of the husk, and feeding on it with voracity.
But for every cherry, apple or ear of corn thus destroyed, a thousand injurious insects are annually
eaten; and so the farmer when he tells his boys "to
show no mercy to the red-heads in the orchard" is
only "saving at the spigot to lose later on at the
bung."

Professor Forbes of the Illinois University examined the stomachs of a number of these birds during
the month of May and found that their food at that
season consisted of: canker-worms, 15 per cent.;
beetles and other injurious insects, 65 per cent.; seeds
and grain, 20 per cent.; thus proving their value as
insect destroyers. In 101 stomachs examined at
Washington, 50 per cent. of the food was of animal,
47 per cent. of vegetable, and the remainder of mineral origin. Beetles and grasshoppers formed 36 per
cent. of all the food, while the vegetable portion was
mainly wild fruits, though 17 of the stomachs contained corn.

The red-head is ever ready to pick a quarrel, not
only among members of his own species, but frequently with other birds, and their noisy chatter when
thus engaged too often makes a bedlam of our fairest
woodlands and drowns out the pleasing notes of the
more favored songsters.

18

V.

'Twas the day before Christmas—bright and beautiful—a winter day in name only. As I tramped towards
the boundaries of one of nature's domains, far beyond
the city limits,

A balmy thermal breeze
To me,
From sunny, southern seas
Came joyously.

Ah, rare the day
At Christmas tide,
When on such breeze
One's thoughts can ride.

Reaching the border of a thicket enclosed by an old
worm fence, I seated myself on the topmost rail of
one of its panels and waited for sound or sight of animal life. I had not long to wait; for soon the chatter
of a noisy jay came from the top of a near-by oak,
while a crow in the midst of the thicket began to
challenge me with his limited vocabulary of "*hah-hah,
hah-hah.*" His call was immediately answered by one
of his brethren in a neighboring woods—this one by
still another farther away, and thus a series of signals
traveled across the township—a wave of crow laugh-
ter, as it were—of which I, perhaps, was the involuntary cause.

Suddenly another sound broke upon my ear, a clear,
ringing, highly musical note—"*willy-way, willy-way,
willy-way,*"—repeated at short intervals. It resembled
somewhat the whistle of the cardinal grosbeak, or
"red bird," but was a far louder and more pleasing

call. I was thrilled with delight, for I recognized instantly the spring note of an old acquaintance, the

The Great Carolina Wren.

great Carolina wren. Few, indeed, are the winter days when this, the characteristic, vernal love call is uttered; but the "sunny, southern" breeze had brought joy also to the wren's heart, and forgetful of season—forgetful of the many cold, dreary days which must intervene before the call could be rightfully issued—conscious only of the warm and soothing sunshine, the singer made the welkin ring for rods around.

The wren was not visible, and was much farther away than I thought, when I started toward the seeming source of the song. The resonant quality of its note caused, as it were, the whole copse to ring with the sound, thus deadening, to a certain extent, the direction from whence it came.

Fig. 77—Great Carolina Wren. (After Coues.)

After some ten minutes' search, during which the notes were stopped a number of times and then re-

sumed, I espied the songster sitting atilt of a splinter
which extended upward from the center of the stump
of a freshly cut tree. His tail was bent downward,
his head raised heavenward, while the rich rolling
notes were issuing forth with a force and energy
seemingly wholly at variance with the size of the
bird.

Unfortunately he saw me too, and instantly the call
was hushed, and with a headlong dive he went into
the depths of an adjacent brush heap. I stepped
behind a tree and waited, for the Carolina wren shares
the restlessness and prying curiosity of all his tribe,
being continually on the go—fidgety—all starts and
jerks—like some people we know in whom the nervous tissue largely predominates.

Sure enough he soon appeared on the side of the
brush pile nearest me, peering from among the dead
leaves with an inquisitive air, jumping from twig to
twig of the brush—all the time teetering his body in
a peculiar wrennish fashion, and performing other
odd, nervous antics as if possessed with the very spirit
of unrest. Finally he reached the topmost twig of
the pile of brush from which commanding situation
he resented my intrusion upon his domain with a
series of scolds—*prr*—*prr*—*prr*—for twenty or more
times, emphasizing each utterance with a quivering
shake of his whole body and especially of his tail.
Then another dive into the brush—a flitter and a
flutter close to the earth—and he was hopping beneath
a fallen beech which, a few rods away, was supported
on its branches some feet above the ground. There
let us leave him for a short time while we make known

some of the more important facts concerning his life history.

The great Carolina wren is "great" only in comparison with other wrens, being the largest of six species which occur in Indiana, and the only one which spends the entire year with us. Another, the little winter wren, nests in northern regions and spends the cold season from the latitude of central Indiana, southward. The other four are "summer residents," nesting in the State and going south in autumn. All are plainly colored, being chiefly brown; and all, in the words of that eminent ornithologist, Elliot Coues, are: "Sprightly, fearless, and impudent little creatures, apt to show bad temper when they fancy themselves aggrieved by cats or people, or anything else that is big or unpleasant to them. They quarrel a good deal and are particularly spiteful towards martins and swallows whose homes they often invade and occupy. Their song is bright and hearty and they are fond of their own music, but when disturbed at it they make a great ado with noisy scolding."

The bill of the Carolina wren is stouter than that of the other species, and its front half is somewhat decurved or bent downwards. The total length of the bird is $5\frac{1}{2}$ to 6 inches, the tail being a little shorter than the wings. In color it is clear reddish-brown above, brightest on the rump, and with the feathers of the wings and tail finely and prettily barred with blackish. Below it is a tawny yellow. The throat and cheeks are gray, while a narrow whitish line curves backward over the eye, and a bar of small white spots crosses the wings near their base.

Other common names by which it is known are "mocking wren" and "large wood wren," while its long Latin name is *Thryothorus ludovicianus* (Gmelin), the first part of which is derived from two Greek words meaning "reed" and "leaping," while the last or specific name is the Latin for "of Louisiana," from which region the bird was first described.

Its chosen haunts are the wooded or rocky banks of streams, piles of logs and brush heaps in clearings, or the zigzag lines of the old rail fences with their corners full of bushy shrubs and fallen weeds; indeed, wherever nature, accident, or design has provided a place where it can make itself conspicuous one instant and be entirely concealed the next. There, too, hidden beneath lichens, in the depths of fungi, or in the cracks and crannies of rail or log, its favorite food, spiders, ants and gnats abound. Its thick-set, bulky body, short wings and tail, and slender, slightly curved bill, are especially adapted to an insectivorous life close to the ground. The wren, therefore, is seldom if ever seen in the tops of tall trees, but sometimes ascends their trunks for quite a distance, peering beneath every piece of loose bark and entering every knothole through which it can squeeze its body, in search of its esteemed spider diet.

The nest of this bird is a bulky structure, composed of strips of bark and corn stalks, grasses, leaves and fibrous roots, and usually lined with feathers, corn silk, or horse hairs. It is placed in any odd nook or cranny that its owner fancies, such as the cavity of a log or stump, the angle of a fence between the lower rails, in a pile of logs or brush—anywhere, in fact, that offers a snug retreat. Often it is arched over at the top, the entrance being at one side and only large enough to admit the builder. The eggs are five to seven in number, reddish-white, thickly spotted with various shades of brown, and measure .73 x .60 of an inch. There are two, sometimes three, broods each season, but many of the young fall a prey to carnivorous snakes, weasels and small owls.

In winter the sexes separate; each pre-empts for itself a certain territory as a forage field, and woe betide any feathered form of moderate size which ventures upon its chosen domain. As Dr. Abbott has well said : "The tenants of the wild woods know the wrens full well and usually give them a wide berth. They realize that they are petty tyrants, suffering no intrusion and excusing no blunders; particularly so when something has gone wrong with them; then it is a word and a blow, and the blow first. Even the hornets stand back when there is a riot in wrendom."

Thus the winter days of our wren are spent in spider hunting, fighting and singing; for besides the spring call and scolding chirps mentioned above, he has many other notes, some of which are as varied and pleasing as those of the brown thrush. The nights are passed in hollow rail or limb of tree, and when the face of earth is clad in snow and ice probably the major part of many a day is spent there too, in fasting and musing—if a bird can muse—o'er the victories won in the past and the battles to be fought in the future.

* *

Wishing to reward the one I had seen with a

Christmas repast, I turned over a large log, on the
under side of which were a colony of ants and a number of fine fat grubs; then, slipping around the fallen
beech under whose limbs he had dodged, I started
him towards the uncovered feast. After some maneu-
vering he reached the log, espied the menu—and then,
how he fed! Turkey, mince pie, cranberry sauce, and
all, those grubs and ants were to him, and thus, in
part, I paid him back for his morning song.

VI.

The subject of the present sketch, the winter wren,
is one of the smallest, and to my mind one of the
most interesting of the fifty or more species which
pass the cold season in this latitude. The diminutive
size of the bird, its quick motions, and especially its
brown color, resembling so closely that of the dead
leaves and grass among which it flits, cause it to

**The
Winter Wren.** remain unseen, unsuspected, and often
quite unknown to the ordinary rambler
in the woods. Even the naturalist,
with trained eye and ear ever open to listen to nature's
sounds, counts that day a fortunate one when he
catches a glimpse of the little creature as it hops or
flits close to the ground, in and out of a fence corner,
or from side to side of a brush pile or log heap. For
it possesses the wrennish peculiarity of being ever on
the go, and although it rarely uses its wings except
for a short flutter from one bush or angle of fence to
another, yet it hops slyly and rapidly about, appearing
perhaps for an instant, then suddenly lost to view.

When thus engaged in active search for insect life, and I never saw one of them otherwise, its bit of a tail is somewhat spread out and raised almost perpendicularly over the back; the neck is bent forward and the straight bill stuck out ahead; thus giving the little bird a most determined and inquisitive air.

The length of the winter wren is not over four inches from point of bill to tip of tail. The bill is very straight, slender and conical. The tail is shorter than the wings, which reach to its middle. In color the body is deep reddish-brown above; the brown being everywhere, except on head and middle of back, transversely barred with dusky. Below it is pale reddish brown, sharply barred on the posterior half with dusky. A line over the eye, some obscure streaks on the sides of head and neck, and some bars on the outer wing feathers are whitish.

Other common names, besides "winter wren," are "bunty wren" and "little log wren," while the long Latin name is *Troglodytes hiemalis* Vieillot, the first part of which means "cave dweller" and the last "wintry."

Fig. 78—Winter Wren. (After Coues.)

Within the past five years I have seen twenty or more specimens of this little wren between the months of October and April, yet I have never heard one of

them utter other sound than an occasional chuckle
like the syllables "*puiq-quap*," or a low wren-like
chirr, when startled or frightened. They are most
common in Indiana during the spring and fall migrations, as large numbers of them spend the winter
farther southward, and must, therefore, pass through
this region on the way to and from their summer
homes. They nest in the pine forests of Wisconsin,
Michigan and British America, especially in damp,
swampy regions, where the ground is covered with
fallen trees and logs, piled upon one another and
covered with rich moss.

·In such a place the male is said to sing most exqui-
sitely, one author describing its song as: "Very lively
and hurried, the notes seeming to tumble over one
another in the energy with which they are poured out.
They are full of power, though many are shrill, and
are garnished with many a gay trill; in some passages
reminding one of the Canary bird's song, though
infinitely finer." Audubon, that prince of American
ornithologists, appears also to have been enchanted
with the song as the following words will testify:
"The song of the little winter wren excels that of any
bird of its size with which I am acquainted. It is
truly musical, full of cadence, energetic and melodious;
its very continuance is surprising, and dull indeed
must be the ear that thrills not on hearing it."

The nest is placed in cavities about the roots of
stumps or in the tangled piles of fallen trees and
branches. It is composed of small twigs interwoven
with moss and lichens and lined with the feathers of
other birds or with rabbit hair which the bird has

picked up in the vicinity of its home. The eggs are five or six in number, pure crystal white, spotted with bright reddish-brown, and measure .67 x .48 of an inch.

During the winter, while the woodpeckers, nut-hatches, titmice and chickadees are fast lessening the hordes of insects which inhabit the trunks and limbs of trees, the wrens are doing the same good work among the logs and stumps close to the ground. There they have less competition, and so find the struggle for existence less deadly than they would higher up among their larger arboreal kin. Thus, in the course of time, each form of bird has found for itself that place in the realm of nature best suited to its existence; and there, most often, do we find it, ever on the search for its favorite food and ever on the alert to prevent itself being eaten by some animal higher in the scale of life.

VII.

The avian or bird fauna of Indiana has been modified in many respects by the presence of the white man and his progressive civilization. The Carolina paroquet and ivory-billed woodpecker, once frequent in our forests, have receded before his advance, and, like the Indian, buffalo, bear, elk and deer, are gone forever. Only in the densely wooded districts of the southern and south-western States do these two noble birds still exist. Even there their numbers are constantly lessening, and, in the opinion of the best judges, a quarter of a century hence they will be known only in museums and in literature. The pileated wood-pecker or log-cock, once abundant and second only to the ivory-bill in size and beauty, is also nearly ex-tinct in this State, occurring in small numbers only in the densely wooded portions of the southern coun-ties. Wild pigeons, formerly so numerous that flocks of them were visible in the air for hours at a time, are no longer seen; while wild turkeys, once the source of many a day's exciting sport, have almost wholly disappeared. Many species of hawks and owls are becoming scarce on account of the unmitigated and senseless warfare waged against them by persons wholly ignorant of the good these birds do in preying upon the hordes of smaller mammals and larger in-jurious insects.

On the other hand, that well known bird, the quail or bob-white, was probably absent or confined to but few localities in the State at the time of its first settle-ment, and has steadily increased in numbers as the forest has been cleared away. The lark-finch, a hand-some member of the sparrow tribe, has made its way in from the plains and prairies of the west and has spread eastward as far as the Alleghanies. Its con-gener, the dick-cissel or black-throated bunting—once rare—now rolls forth at June time its characteristic warble from the borders of every clover or grain field in the State; while that pestiferous alien, the English sparrow, an enforced immigrant from good Great Britain, has multiplied by countless thousands within the past decade.

Many other instances of decrease or increase in the number of birds, caused either directly or indirectly

by the settlement of the country, could be given, but
let us now note a few of the changes in habits which our
native birds have adopted since the white man came—
and so lead up to one discovery in particular which
the shrikes or butcher birds have made and put to use.

Fig. 79—Dick-cissel or Black-throated Bunting. (After Judd.)

The chimney swifts, before the advent of civilized
man, nested and roosted in hollow trees, but these
trees becoming scarce and chimneys frequent, and
possibly more to their liking, they in time forsook,
almost wholly, the former for the latter.

Woodpeckers, before the settlement of the prairies, were confined to forest areas on account of a lack of suitable nesting places; but they have discovered that the cornices of buildings, church steeples, telegraph poles and even rotten fence posts, will, with a little labor, furnish a lodging place for eggs, and so they have spread far and wide over the treeless regions of the western States. Bank swallows have utilized for nesting places the artificial roadbeds of railways in some of the flatter counties of the State, and hence their presence in such regions does not antedate that of the "iron horse."

Many rapacious birds, as the smaller hawks, owls and kingfishers, since the clearing away of the forests, use the top of telegraph poles as resting places from whence they can swoop down upon such unfortunate prey as may come within their vision; while blue birds and swallows, in lieu of a better resting place, often line themselves along the wire and twitter and chirp to one another, wholly unconscious of that electric force which is propelling the thoughts of man at lightning speed along the slender thread beneath their feet.

But the shrikes or butcher birds have put to use another device of man, and in a peculiar manner: namely, the barbs upon the barbed wire fences as spikes upon which to impale their prey. For the shrikes are the bushwhackers among birds. No others are so notorious for cruelty and rapacity. Not only for food, but apparently for the gratification of a blood-thirsty instinct, they kill forms of life beneath them merely as a means of killing time. They alone of all birds impale their victims upon sharp pointed projections. Their food consists of mice, small birds, snakes, beetles and grasshoppers. Formerly these birds visited only low-ground thickets where crab-apple, haw-thorn and honey-locust or "thorn-trees" abounded, upon whose sharp twigs and thorns they hung their victims. But since the advent of the barbed wire fences the shrikes have appeared everywhere along upland fields, finding in the sharp, stiff barbs just the kind of an impaling spike they wish. Why the prey is thus hung on thorn or barb has not, as yet, been satisfactorily explained, for it seems that objects so impaled are afterwards seldom touched by the bird.

Two species of shrikes inhabit Indiana. One, the logger-head shrike, *Lanius ludovicianus* L., is a summer resident, arriving from the south about April 1st and departing thither about mid-October. This is the species which impales so many grasshoppers and beetles along the wire fences during the summer and autumn. On one October day I gathered fully a pint of such impaled insects from a fence row half a mile long, and found that they represented sixteen species; eight of grasshoppers, two of katydids, and six of beetles, all injurious, so that this bird, although savage and bloodthirsty, is of great benefit to the farmer and fruit grower.

No sooner has the logger-head departed for the south, than its cousin, the great northern shrike, **The Great Northern Shrike.** arrives from the north to spend the winter with us. In general appearance this latter species closely resembles that exquisite singer, the southern mocking bird;

but the shrike is readily distinguished by its more
bulky form and its much stronger hooked and notched
bill. In size it is about the same as the common
robin, measuring 9½ inches from tip of bill to end of
tail. Above, the general color is a clear bluish ash,
somewhat paler on the rump. Below, it is a dirty
white, everywhere crossed with fine, wavy blackish
lines. The quills of wing and tail and a broad bar
along the side of head are black: while a white spot
is situated upon the lower half of each wing.

Fig. 80—Great Northern Shrike. (After Coues.)

The northern
shrike nests
only in British
America, but in
winter ranges
southward to
about latitude
36 degrees. Its
scientific name
is *Lanius borealis* Vieillot, from two Latin words
meaning "butcher" and "of the north." During
severe winters these birds often appear about the
suburbs of cities and prey upon the English sparrows,
and sometimes become so bold as to fly into open
windows and attack a canary, even in the presence of
human beings. By the Germans they are often called
neuntödter or "nine killers," from the belief that they
catch and hang up nine mice or nine sparrows each
day.

They are treacherous birds and use many devices to
get within striking distance of their prey. Dr. Abbott
has well described their actions as follows : " I remem-

ber one, demure as a scheming crow, with eyes half
shut and with not a trace of treachery or cunning in
his face. His blue and white plumage, tastefully
trimmed with black, made him conspicuous, but he
lessened the ill effects of this by the manner he assumed.
No bird, however timid, would step aside for such as
he. Indeed, they perched upon the same branch of
the tree he was on, almost upon the same twig, and—
where was he?

"Like a flash the shrike had disappeared, and now,
fifty paces distant, he is perched upon another tree,
plucking feathers from a kinglet's head and regaling
himself with his victim's brains."

Shrikes seldom sing. On a sunny December day,
I was, however, favored by the song of a butcher
bird. I was on the lookout for him for I had found
a sharp-nosed shrew-mouse hanging on a barbed wire
fence, and knew that a shrike had its winter quarters
in the immediate vicinity. Suddenly a bull-headed
bird, with chops muffled in black, sprang from a fallen
thorn tree to a wild cherry in a near-by corner of the
rail fence, and I knew that the butcher was in my
presence. Eyeing me furtively for some time he
jumped from branch to branch towards the top until
finally he reached the uppermost twig when suddenly
he uttered four sharp notes. They were much like
the warning cry of a cock when he discovers a hawk
or large bird in the air—"*crr—crr—crr—crr.*" Then,
after an interval of a few seconds, he began a song
which was continued for at least five minutes. It was
uttered in a joyful manner but at the most was a
monotonous ditty, a "*puit-tuit-toot-e-ree, puit,*" etc.

During the singing he gazed attentively at me as if seeking to note the effect of his music, taking no notice of the wrens and sparrows which were flitting about in the fallen thorn. As I stepped towards the tree on which he was perched he uttered a harsh note, a shriek of protest, as it were, at my intrusion; then giving two or three vigorous strokes with the wings, he folded these organs close against his body, and, with a peculiar gliding motion, passed swiftly into a bushy oak shrub a hundred yards away.

VIII.

The student of nature soon learns to notice the interdependence existing among all forms of living objects. All animals are wholly dependent upon plants for their existence, for plants alone can change inorganic matter—earth, air and water—into starch, sugar and other organic food-stuffs for animals. Plants alone can collect and store up in these foods the radiant energy of the sun's heat and light, and so transmit it to the animals within whose bodies it is changed into animal force and used as nervous, muscular and gland power to perform the duties of animal life.

On the other hand plants receive many benefits from animals. Insects carry the pollen from flower to flower and so aid in the fertilization and cross fertilization of the plants. To bring about the visitation of insects to the reproductive parts or anthers on which the pollen is produced, the plants have developed honey secreting glands about the base of the anthers

and then surrounded them by showy petals so that
the insects may readily find their way to the honey
and in their search for it unconsciously scatter the
pollen. In this way all the showy parts of flowers
have originated.

Again, birds and plants are mutually dependent,
the birds feeding upon the fruits of the plants and in
turn scattering or distributing the indigestible seeds
far and wide over the face of the earth. Indeed, all
the fleshy parts of fruits have in time been developed
around the seeds for the sole purpose of bringing
about the distribution of the latter. Man gathers
apples, oranges, or apricots from the tree for the
sake of the flesh or pulp, not for the seeds; but in
getting the pulp he carries the seed far away from
the parent tree. Cherries, raspberries, strawberries
and many of the fruits of wild plants are very attractive to the palate of birds, and the latter are therefore the chief agents in the distribution of these fruit
producing plants.

Among our frugivorous birds the cedar bird or
cherry bird ranks pre-eminent for the great variety
of wild fruits which it eats. This bird is a permanent resident in Indiana, but is most abundant during
the spring and fall migrations as many of them spend
the winter farther southward. It is almost always
seen in flocks, usually from forty to a hundred together.
It has no song and no gaudy colors, yet from the

The Cedar Bird. delicacy and softness of its plumage it
is one of our most beautiful birds; and
during the craze for feather ornamen-
tation, which was so prevalent a few years ago, and

which still exists among certain stages of society, the cedar bird was one of the most common species seen on women's hats.

In length it measures about seven inches from point of bill to tip of tail. Above, the general color is a cinnamon-brown, paling to slaty ash on the wings and

Fig. 81—Cedar Bird. (After Beal.)

tail, the feathers of the latter being tipped with a band of yellow. A broad stripe of black passes from the nostrils back along the side of the head, and the chin and forehead are also black. The tips of some of the wing feathers, and sometimes also of the tail feathers, bear little oval, flat, leaf-like appendages which resemble red sealing wax; while the head is ornamented with a conspicuous crest which can be raised or lowered at pleasure.

The wax-like tips on the wing feathers are only prolongations of the shafts with curiously formed pigment cells containing an abundance of red and yellow coloring matter. Their presence gives rise to two common names, other than those given above, by which the bird is known, namely: cedar wax-wing and Carolina wax-wing. The scientific name is *Ampelis cedrorum* (Vieillot), meaning "fruit eater," and "of the cedars;" as one of the favorite foods of the bird is the cedar or juniper berries so common in some localities.

The cedar bird, like many human beings, lives to eat. Although an abundant species throughout the State it is so capricious in its movements that its presence or absence in a certain locality appears to bear no relation to season or weather, the question of food supply alone being probably the controlling influence in its wanderings. It feeds on cherries, both wild and cultivated—whence its name, "cherry bird"—on the berries of the sour gum, dogwood, bitter-sweet and pokeweed, and will often so gorge itself with these as to be almost unable to fly. Whenever a flock alights in a tree bearing their chosen food they sit for a time motionless and erect like parrots; then, by a movement of the head, each one takes a survey of his immediate surroundings, after which, one by one, they proceed to the chief business of their lives.

In spring and summer, before the ripening of the fruit and berries, this bird is of great benefit to the farmer, as it then devotes itself almost entirely to catching insects. It is an expert fly catcher, and may often be seen perched upon a dead twig near the top of some tall tree from which it makes its graceful and successful flights after the different insects passing near. In this way it repays more than twenty fold for the cherries it later on devours.

The cedar birds are noted for their extreme sociability and even fondness for their kind. They are among the few birds which appear to be permanently gregarious, i. e., always found in flocks. They build their nests very late in the season, sometimes not till the middle of July, and are seen in flocks up to that time. Several nests are often constructed in the same tree and usually all those of the same flock are built within the compass of a few rods. The nest is a bulky structure composed of many materials, such as bark, roots, twigs, paper, rags and twine, and lined with the finer grasses, hair and wool. The eggs are three to six in number and are slate brown marked with many purple or dark brown blotches.

The sociability of these birds is kept up during the five or six weeks that they are held in one locality by the care of their young and when the latter are ready to leave the nest they remain with their parents.

On a recent January morn my attention was attracted to a flock of birds which was continually flying from some trees to the margin of a small pond and back again. I moved slowly towards them and found them to be cedar birds which were feeding upon the fruit of the hackberry, *Celtis occidentalis* L. They were working rapidly and tearing at the berries so eagerly that as many fell to the ground as were eaten.

About every five minutes a dozen or more members of the flock would sweep in an easy and undulating but swift flight to the margin of the pond and after drinking would fly again to their feast. They made no sound except an occasional lisping *tsip* in a low tone. Occasionally one would be overcome with curiosity and would fly to the branch of the neighboring tree under which I stood and peer down at me, moving his head from side to side while his eyes seemed to sparkle with excitement; then back he would go again, showing as he flew the waxen tips on his primaries and the bright yellow border on his tail feathers.

I envied them their food, plucked from the tree on which it grew, and therefore free from adulteration of any kind; sweet, resembling a black haw in taste; the only drawback being that the pulp or nutritious part is very small in proportion to the bulk of the seed. But the former is sufficient in quality and quantity to attract the birds and therefore serve the purpose of the hackberry; and perhaps many a *Celtis* will owe its future existence to the visitation of the flock of cedar birds seen by me; the seed being dropped in some distant place where its chances for life and growth will be a thousand fold greater than if it had fallen to the earth beneath the parent tree.

IX.

Many of the natural haunts of our winter birds, which formerly existed within easy reach of the city, have disappeared. As a consequence the birds themselves are yearly becoming less frequent in number.

I have in mind a wooded tract, of probably 100 acres, which a few years ago was thickly grown up to under-brush and contained many fallen trees. There, on any day in winter, one could find twenty or more species of birds, on the trees, in the brush piles, or on the ground, each one seeking its food according to its adaptations. Not only birds but many other forms

Fig. 82—Brown Thrush. (After Judd.)

of living things, such as snails, reptiles, batrachians, and insects of varied size and shape, found a congenial home within the borders of those woods. But, although they belong to a wealthy family who could well afford to have spared them, to-day all is changed.

Every brush pile where in summer nested the brown thrush and Carolina wren; every log beneath which the plodding snail, spotted salamander or white-footed mouse found shelter; every shrub and weed on whose leaves and flowers beetles and butterflies of brilliant hue had been wont to feed, are cleared away. The broad and level sward with its unbroken carpet of Kentucky blue-grass no doubt presents an attractive appearance to the eyes of the rich owner; but the tenants of yore, which to him were unknown or despised, though some of them for many years had been his helpful friends—what of them?

To my mind they were the rightful owners of the land. Back in preglacial times, before the overflow from a giant, melting, bulk of ice had carved out the broad valley to the westward—back then, and even before—the first insects, birds and mammals "entered" this tract of land and began on it a struggle for existence. There, for year on year and century on century, has this struggle continued, and though the home and happiness of many of the contestants were destroyed by the clearing up of the under-brush, yet it still continues. Nor will it wholly end until the advancing city will have encroached upon the bounds of this domain of nature, and then man, proud, artificial, unnatural, will forever drive out the rightful denizens and prove himself the fittest in the struggle.

The above is but one example of a thousand going on everywhere about us. To produce that which will bring him wealth—although his coffers may be full to overflowing—man willingly and thoughtlessly causes the death or disappearance of manifold forms of living things and creates sad havoc with the true beauty of nature's own. Like Thoreau, I exclaim: "Thank God, he can not cut down the clouds."

Among our winter birds none was formerly more familiar than the so-called "snow-bird" or slate colored junco. It arrives from the north about October the 15th and its coming is always a precursor of the winter that is to be. During the pleasant weather, when the earth is bare, it seeks the shelter of such wildwood tangles as the one above mentioned. There,

The Snow-bird. scattered among the leaves and on the dead stems of low grasses and weeds, it finds its favorite food, the seeds of wild plants, which it occasionally varies with such small beetles or grubs as it may happen upon. But when such food is hidden beneath a coating of ice and snow the junco has to retreat from its thicket stronghold and then it comes trooping about the dooryards and barnyards of man, ever ready to pick up those "crumbs of comfort" which are dropped intentionally or otherwise in such places.

The snow-bird belongs to the great family of *Fringillidæ* which comprises the sparrows and finches, 38 of which are known to occur in Indiana, 17 of them being found here in winter. The chief character which distinguishes this family is a thick cone-shaped bill which is shorter than the head and abruptly angulated or drawn down at the corners of the mouth. With this they can crush the hard outer shell of most of the smaller seeds and feed upon the rich, nutritious kernels within. The English sparrow and Canary bird are two familiar members of the family.

Although seeds form the main diet of all these birds, yet, in early spring and summer when seeds are scarce, they turn to insect life to furnish them sustenance;

and **Prof. S. A.** Forbes found that 91 per cent. of the food of 47 sparrows which he killed in an orchard in May, was composed of insects; four per cent. being canker worms, which are so injurious to the foliage of the apple tree. Moreover, the young of the 17 species which nest in the State are fed wholly upon insect food, so that, all in all, the family is a most beneficial o n e to our husband-men.

In size the snow-bird is below the me-dium, measur-ing but 6¼ inch-es in length. In color it is a uni-form ashy or blackish g r a y a b o v e, some-what darker on the head. Be-low, all back of the breast is pure white, as

Fig. 83—Snow-bird. (After Coues.)

are also the two or three outer feathers of the tail. These feathers are ever a sign of its identity, for as it flies it spreads its tail enough to show their edges. No other bird of similar size possesses them except the grass-finch or vesper sparrow, which is everywhere streaked, both above and below, with reddish-brown and dusky.

From the above description it will be seen that the name "snow-bird," by which this sparrow is so commonly known, is a misnomer as far as color is concerned, and was probably given to it on account of its habit of flocking about houses and barns after every snow storm.

Its scientific name is *Junco hiemalis* (L.), from two Latin words meaning "a rush" and "wintry." The common name, "slate-colored junco," which, among naturalists, is coming into general use, is sufficient to properly distinguish it from other members of the same genus.

While the snow-bird is known to every one in the country as a common and familiar winter resident, there are few people but to whom its coming and going is a mystery, and the question is often asked: "What becomes of the snow-birds in summer, and where do they nest?" Many are the answers given to this question by persons who know little or nothing of the habits of birds. A common belief, and one which was upheld by an article which appeared less than a dozen years ago in one of the leading newspapers of this State, is, that our common sparrows, such as the field sparrow and grass-finch, change color in fall, becoming snow-birds, which they remain until spring, when they don their other dress and again become sparrows.

The question is easily answered by any one who will give the matter a little thought. The snow-bird nests from Michigan and Wisconsin northward, and in autumn many of them stop with us to spend the winter instead of all going farther south as is the case with so many birds. This makes it a true "winter resident," of which, among the sparrows, but five occur in Indiana.

While here the snow-bird is always seen in small bevies or family groups, never singly. A true brotherly and sisterly love seems to animate these groups. Their members are never seen quarreling among themselves as is so common an occurrence with their pesky cousins, the English sparrows; but when one is crippled or ailing all the others vie with each other in carrying food to it and giving it every needed attention. Their ordinary note is a short, sharp, emphatic *"chip"*, rapidly repeated as the bird is flushed; but in the spring, as the days become warmer, they delight to sit in the low branches of trees and sing a very sweet, suppressed song, as if tuning up in anticipation of the coming mating in that far northern country for which they will soon depart. All in all, though dull in color and lacking in brilliant song, these little snow-birds have many charming habits, well worthy the study of any one interested in our feathered friends. Were they forever taken from our midst we would sadly miss them on those days when murky clouds o'erspread the sky and snow and ice enshroud the lap of earth.

"Better far, ah yes! than no bird
Is the ever-present snow-bird;
Gayly tripping, dainty creature,
Where the snow hides every feature;
Covers fences, field and tree,
Clothes in white all things but thee;
Restless, twittering, trusty snow-bird,
Lighter heart than thine has no bird."

X.

While tramping through woodland, field and meadow in search of "first hand" knowledge, I often think of the many riches possessed by the farmer's son which he wots not of. His father's fields have thousands of tenants which he never sees. In their proper season wild flowers of brilliant hue and delightful odor bloom all about him yet are passed unnoticed. Every corner of the old Virginia rail fences holds countless treasures of brilliantly colored insects, yet he knows only the six or eight species of homely ones which are especially injurious to his father's crops. In the proper season the orchards and woodlands on the old homestead are full of sweet singing warblers and vireos whose notes and plumage may be, for the time being, all his own; yet he sees and hears them not. The rainbow darter and its cousin, the green-sided darter, swim up and down the ripples of the brook which flows through the wood's pasture of his country home, yet the sunlight which they reflect from their gilded sides ne'er strikes the eyes of the farmer's boy. If I possess treasures I wish to know it and not pass my life surrounded by them and yet in continual ignorance of their presence.

How many of my readers, for example, whether reared in country or city, have ever seen a cross-bill alive? I lived upon a farm until I was of age and did not know that such a bird existed although it was probably found every winter within a mile of my home.

Like the snow-bird, the cross-bill belongs to the great family of *Fringillidæ*, whose members are commonly known as finches and sparrows—all having a

The American Cross-bill.

thick cone-shaped beak for cracking seeds. But the beak of the cross-bill has, in time, undergone a wonderful change, or, in other words, has become adapted to the habits of the bird. For, instead of the two mandibles

Fig. 84—American Cross-bills. (After Coues.)

meeting on a level as in other finches, the upper curves down to the right of the lower, which at the same time curves upward. In this way they partly cross one another, thus giving rise to the common name of the bird.

The bill is thus fashioned to extract the seeds of pines and other similar trees from the cones, and the cross-bills, by the great strength of the muscles of the jaw and these strong, oppositely curved mandibles, are able to pry open the tightly appressed scales of the cones and extract at pleasure the nutritious oily seeds. Other birds are equally fond of these seeds, but have to wait for the alternate thawing and freezing of spring to loosen the scales.

In size the cross-bill is somewhat larger than the English sparrow, measuring about 6½ inches in length. The wings are very long and pointed, reaching beyond the middle of the narrow, forked tail. The color of the bird varies greatly according to sex and age. The old males are brick-red, darkest across the back, and have the wings and tail a uniform blackish-brown. The females of all ages are dull greenish-olive with a yellowish tinge on the crown and rump; while the young males are a curious mixture of brick-red, greenish-olive, and yellowish.

There are two species of cross-bills occurring in Indiana, and the above description applies to the most common one which is known by the name of the American or red cross-bill. The other species is the white-winged cross-bill and may be readily distinguished by the presence of two conspicuous white bars across each wing. It has the same habits as the one above described, but has been noted in the State only on a few occasions.

The scientific name of the American cross-bill is *Loxia curvirostra* L., from two Latin words meaning "crooked" and "curve-bill"; while that of the white-winged species is *Loxia leucoptera* Gmelin, the word "leucoptera" meaning "white-wing." Thus the scientific name of each kind of plant or animal is often based upon some important and noticeable character possessed by the species.

Perhaps no birds are more erratic in their movements than the cross-bills. They appear and disappear from a given locality in the most unexpected manner. In Indiana they are most likely to be found during severe weather in January and February. They go in flocks of from six to forty individuals, usually flying high in air and for great distances at a time. While on the wing they keep up, almost continually, a loud clear call note, which is quite similar to that made by a young chicken in distress. Always alighting in the tops of pines or other evergreen trees, each individual chooses a cone and immediately begins to extract the seeds.

While thus feeding they are extremely gentle and social, easily approached, and may even be knocked down with sticks. In the old barbarous collecting days I have shot as many as five or six from a single tree without causing the remainder of the flock to take flight. The two species are alike in all their habits, climbing from cone to cone like parrots, head down or head up at will; twittering as they feed like many other sparrows; and finally, having eaten their fill, with one impulse they hurry out of sight, to be gone, it may be, until another year rolls by.

They nest throughout the coniferous forests of the northern United States and Canada and in the mountains of the southern States, notably in North Carolina and Tennessee. A few of the young of the previous season either remain in Indiana throughout the summer, or visit here at that season, for I have

20

taken them in Putnam County in July, but no well authenticated record of their nesting in the State has, as yet, been made.

Among the superstitious many curious legends exist accounting for the origin of certain well known characters of our common birds. For example, the red breast of the robin is said to have resulted from a habit that these birds had, in the misty past, of filling their bills with water which they carried to the brink of Hades and dropped down to the thirsty unfortunates below, their breasts meanwhile becoming scorched by the flames from the infernal regions. In like manner the curved mandibles of the cross-bills are accounted for by saying that these merciful birds tried to pull the nails from the cross, and in so doing twisted their bills in such a way that they will always bear the symbol of their good deed.

XI.

When, after a few weeks of imprisonment within the city, the naturalist goes forth to make new friends among the denizens of the woods and fields, the pure country air has, at times, a curious effect upon his mind, causing strange thoughts to well up therein concerning the relations of man to the animals and plants about him, and especially to the earth itself. Thus, on one of the pleasant afternoons of late February, as I tramped over a wooded knoll east of the city, I found myself likening the earth to a great round animal, moving on an eternal journey through space, and of mankind as mites, preying upon its back, scratching its thick rough hide with their tiny implements of toil and so causing it to yield them sustenance as does the mite or tick the animal it lives upon.

The sight of the smooth lichen-covered bole of a beech caused this revery to vanish and I began to ponder over the power of sunlight, which, after traveling ninety and more millions of miles, had built up the carbon, hydrogen and oxygen, which the roots and leaves had gathered, into thousands of pounds of solid wood for the use of man. Even the green slime or protococcus found on the north side of the tree has its part to bear in the economy of nature. For it is composed of myriads of little cells, each a complete plant in itself and one of nature's disinfecting organs; which, by the aid of the all powerful light of the sun, takes up the impure carbon-dioxide and sets free oxygen, pure and invigorating, for man and beast.

But in the beech were birds, forms of life of which I was primarily in search. Two species there were which in winter are almost inseparable; namely, the tufted titmouse and the black-capped chickadee. Both are permanent residents, that is they are found here at all seasons of the year but appear most abundant in winter; probably because the trees are then bare and the birds can be more readily seen.

Both belong to the same family, the *Paridæ*, of which we have but five species in Indiana; namely, one titmouse, two chickadees, and two

The Tufted Titmouse. nuthatches. Of these the tufted titmouse is the largest and yet it is below the average bird in size, measuring but $6\frac{1}{4}$ inches

in length.   In color it is a uniform leaden-gray above,
except a narrow streak of black across the forehead.
Below, it is a whitish ash, with the sides tinged with
dull reddish-brown.   The feathers of the head are
long, and, when the bird is excited or angry, can be
raised into a conspicuous crest, whence the common
name of "tufted titmouse."

From the black bar across the forehead it is sometimes called the black-front-
ed titmouse, while its Latin
name is *Parus bicolor* L., the
former word meaning "tit-
mouse" and the latter "of
two colors."

In winter perhaps no
bird is more abundant in
the wooded portions of the
southern half of Indiana,
than this species.   Roving
in   restless,   noisy   troops
through the woods, scolding

Fig. 85—Head of Tufted
Titmouse.

at every intruder and calling to one another in harsh
tones, it soon renders itself conspicuous to every one
who is beginning to take an interest in our feathered
fauna.   Its ordinary note is a rather monotonous
"*dee-dee-dee*" often repeated as if from habit.   Its
song is a loud clear whistle resembling the syllables
"*peto-peto-peto*" uttered in a defiant tone, as if chal-
lenging all other birds within the compass of its
voice.   When angry it raises its crest and utters a
series of chirps which appear to be imitations of
the notes of other birds, those of the blue jay being

recognized. While searching for insect food it moves by short sudden leaps and flights from branch to branch, suspending itself readily in all attitudes.

When forest food is scarce it often approaches gardens and orchards, and then only do we see it on or close to the ground, ready to pick or tear at any vegetable or animal food which may be to its liking. In summer these winter bevies separate; and each pair seeks some natural cavity such as a hollow in the fork of a tree or a deserted hole bored by a woodpecker. This is lined with bits of moss, leaves and grass, and in it the eggs, four to six in number, are deposited. In color they are white, sprinkled with reddish-brown and lilac, and measure .75x.56 of an inch.

XII.

The black-capped chickadee is a better known bird than the tufted titmouse, its colors being more striking and its habits more sociable than that species.

It is also much smaller, measuring but five inches from point of beak to tip of tail. The crown, nape, chin and throat are a rich glossy black, and contrast strongly with the grayish ash of the remainder of the body.

Fig. 86—Head of Black-capped Chickadee.

The Black-capped Chickadee.

As stated above, both these birds belong to the same family and even to the same genus, so that the first part of the Latin name, *Parus atricapillus* L., is the same for both. *Atricapillus* means "black haired," and a common name, perhaps as much used as the one mentioned, is "black-capped titmouse." It was of this fluffy little bird in its modest dress of black and gray that Emerson wrote as follows:

> " This poet, though he live apart,
> Moved by his hospitable heart,
> Sped, when I passed his sylvan fort,
> To do the honors of his court,
> As fits a feathered lord of land;
> Flew near, with soft wing grazed my hand,
> Hopped on the bough, then darting low,
> Prints his small impress on the snow,
> Shows feats of his gymnastic play,
> Head downward, clinging to the spray.
>
> Here was this atom in full breath,
> Hurling defiance at vast death;
> This scrap of valor just for play
> Fronts the north-wind in waistcoat gray."

He is indeed a joyful little creature, flitting ever about, hither and thither, clinging to the side of a tree one minute and picking at the moss on a branch the next. His ordinary food consists of the insects which hide in the crevices of bark, spiders' eggs, and, perhaps, the tender buds of trees.

His winter note of "*tche-de-de—de-de*" is the one most commonly known, but in spring it gives way to a pleasing "*phe-be*" which is, perhaps, his vernal love call.

The nest is built in a dead stump or tree in a hole excavated by the bird itself. The eggs are white, sprinkled with reddish brown and measure .58 x .47 of an inch in size.

All the members of the family *Paridæ* delight to wander in company and time and again have I found all five of them within an area of a few square rods. On the beech tree above mentioned, there were, however, but the two of which I have written, "the lordly tomtit, with his jaunty crest; the merry chickadee— the former making the woods ring with his earnest

Fig. 87—Bluebird. (After Beal.)

invitation to ramble therein: *here—here—here!* the latter ever winsome as it chirped, in more subdued tones, *chick-a-dee-dee—dee-dee; winter no terror has for me—for me.*"

On the same afternoon the notes of many other birds were heard; notably the ringing quaver of the bluebird again and again—the first symbol of the approaching springtime which all plants, all animals welcome with marks and notes of joy. It was the beginning of a great awakening which once each year comes to all animate life. A few of the insects, plants, etc., began that day to rub their eyes and endeavor to peep forth to see if a new morning was really beginning to dawn. A few days of such weather and they will begin to call to one another the "good morning" of spring. The matin-song or call was that of the bluebird. The hum of insect, the croak of frog and the clear whistle of the red-shouldered blackbird will soon follow in regular and long accustomed order.

HOW PLANTS AND ANIMALS SPEND THE WINTER.*

One of the greatest problems which each of the living forms about us has had to solve, during the years of its existence on earth, is how best to perpetuate its kind during that cold season which once each year, in our temperate zone, is bound to come. Many are the solutions to this problem. Each form of life has, as it were, solved it best to suit its own peculiar case, and to the earnest student of Nature there is nothing more interesting than to pry into these solutions and note how varied, strange, and wonderful they are.

To fully appreciate some of the facts mentioned below it must be borne in mind that there is no such thing as "spontaneous generation" of life. Every cell is the offspring of a pre-existing cell. Nothing but a living thing can produce a living thing. Hence every weed that next season will spring up and provoke the farmer's ire, and every insect which will then make life almost intolerable for man or beast, exists throughout the winter in some form.

If we begin with some of the lowly plants, such as the fresh-water algæ, or so-called "frog-spittle" of the ponds, and many of the rusts and fungi which are so injurious to crops, we find that they form in autumn "resting spores." These are very small and globular,

* Popular Science Monthly, February, 1897.

one-celled bodies, having a much thicker coat and denser protoplasm or contents than are found in the spores often produced in summer by the same plants, and which are destined for immediate growth. The power of life within these winter resting spores is proof against the severest attacks of frost, and they lie snugly ensconced in the mud at the bottom of pond or stream, or buried beneath the leaves in some sheltered nook, until the south winds of

Rusts and Fungi in Winter. March or April furnish the key to unlock the castle of the ice king. Then the spirit of growth within each spore begins to assert itself once more, and, bursting the walls, the contents soon produce the parent or summer form of the plant with which we are most familiar. Thus the spores which the next season will produce the grape mildew and the red rust of wheat exist throughout the winter—the former within the substance of the fallen grape leaf, the latter within the stubble or about the roots of the last season's wheat plants.

If the grape leaves should be carefully gathered and burned, and the stubble destroyed in like manner, not only would the next season's crop of these two parasitic plant pests be wonderfully lessened, but many injurious insects would at the same time be destroyed.

Higher in the scale of plant life we find the flowering annuals bending all their energies during the summer to produce that peculiar form, the "seed," which is only a little plant boxed up to successfully withstand the rigors of winter. The great sunflower, that grows into a giant in a single season and defies

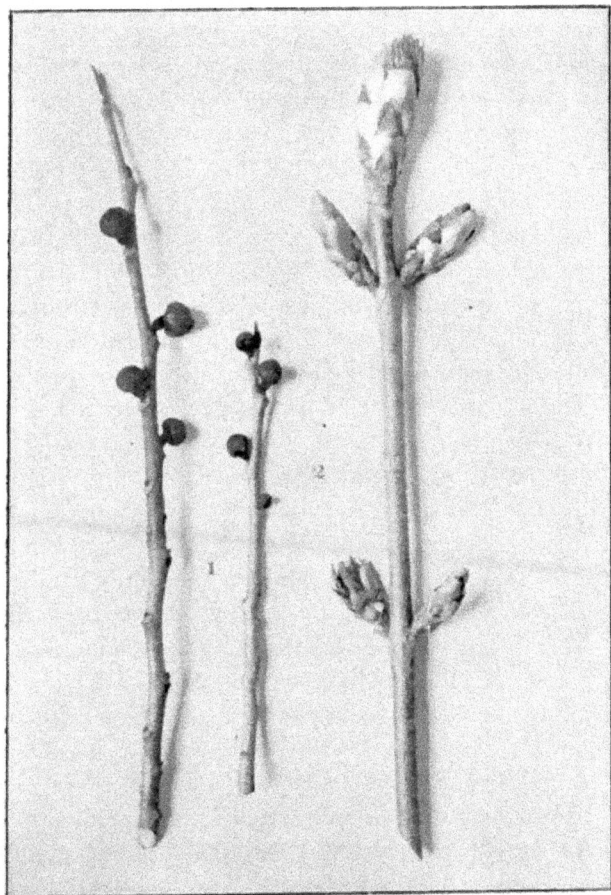

XIV

WINTER BUDS.
1. Papaw.
2. Buckeye.

the summer sun and storm, falls an easy victim to the frosts of autumn. It, however, prepares the way for many successors in the ripened seeds, each one of which, under favorable conditions, will germinate, grow, reproduce its kind, and thus complete another cycle in the realm of vegetable life. The prospective life and activity of a whole field of next summer's waving corn may be considered as stored up in a few pecks of comparatively lifeless seed corn safely housed in the granary. Within its two protecting coats and surrounded by a large store of food, in the form of seed leaf or nucleus, to be used when growth begins again, each little plantlet lives and survives the coldest blasts of King Boreas and his cohorts.

Note, too, the buds and under-ground stems which will furnish the beginning of next season's growth of our biennial and perennial plants. See how they are protected by heavy overcoats in the **Winter Bud Scales.** form of bud scales. Oftentimes, too, as in the hickory and "balm of Gilead" trees, these scales have a coat of resin or gum on the outside to render them waterproof; and some, as those of the papaw, are even fur-lined, or rather fur-covered, with a coating of soft black hairs. Were these protective scales not present, the tender shoots within them, which will furnish the nucleus for next season's foliage, would be seared and withered by the first frost as quickly as though touched with a red-hot iron.

The above are some of the many ways in which our plants, in the course of ages and many changes of environment, have solved the problem of surviving the cold of winter. Moreover, they always prepare

for this cold in time, the resting spores and seeds
being ripened and the bud scales formed over the tender tips of the branches long before the first severe
frost appears.

Let us now take up those higher forms of life-called
animals—"higher," because they are absolutely dependent upon plants for their food—and see how they
pass their time while their food producers, the plants,
are resting.

Beginning with the earth-worms and their kindred,
we find that at the approach of winter they burrow
deep down where the icy breath of the frost never

**Earth-worms
in Winter.**
reaches, and there they live, during
the cold season, a life of comparative
quiet. That they are exceedingly sensitive to warmth, however, may be proven by the fact
that when a warm rain comes some night in February
or March, thawing out the crust of the earth, the next
morning reveals in our dooryards the mouths of hundreds of the pits or burrows of these primitive tillers
of the soil, each surrounded by a little pile of pellets,
the castings of the active artisans of the pits during
the night before.

If we will get up before dawn on such a morning
we can find the worms crawling actively about over
the surface of the ground, but when the first signs of
day appear they seek once more their protective burrows, and only an occasional belated individual serves
as a breakfast for the early birds.

The eyes of these lowly creatures are not visible,
and consist of single special cells scattered among the
epidermal cells of the skin, and connected by means

of a sensory nerve fiber with a little bunch of nervous matter in the body. Such a simple visual apparatus serves them only in distinguishing light from darkness, but this to them is most important knowledge, as it enables them to avoid the surface of the earth by day, when their worst enemies, the birds, are in active search for them.

The fresh-water mussels and snails and the crayfish burrow deep into the mud and silt at the bottom of ponds and streams where they lie motionless during the winter. The land snails, in late autumn, crawl beneath logs, and, burrowing deep into the soft mold, they withdraw far into their shells. Then each one forms with a mucous secretion two thin transparent membranes, one across the opening of the shell and one a little farther within, thus making the interior of the shell perfectly air-tight. There for five or six months he sleeps, free from the pangs of hunger and the blasts of winter, and when the balmy breezes of spring blow up from the south he breaks down and devours the protecting membrane and goes forth with his home on his back to seek fresh leaves for food and to find for himself a mate.

Mussels and Snails in Winter.

Next in the scale come the insects, which comprise four-fifths of all existing animals, and each one of the mighty horde seen in summer has passed the winter in some form. One must look for them in strange places and under many disguises; for they can not migrate, as do the majority of the birds, nor can they live an active life while the source of their food supply, the plants, are inactive.

The majority of those insects which in May or June will be found feeding on the buds or leaves of our trees, or crawling worm-like over Eggs of Insects in Winter. the grass of our lawns, or burrowing beneath the roots of our garden plants, are represented in the winter by the eggs alone. These eggs are deposited in autumn by the mother insect, on or near the object destined to furnish the young, or larvæ, their food. Each egg corresponds to a seed of one of our annual plants; being, like it, but a form of life so fashioned and fitted as to withstand for a long period intense cold; the mother insect, like the summer form of the plant, succumbing to the first severe frost.

Thus myriads of the eggs of grasshoppers are in the early autumn deposited in the ground, in compact masses of forty to sixty each. About mid-April they begin to hatch, and the sprightly little insects, devoid of wings, but otherwise like their parents, begin their life-work of changing grass into flesh.

A comparatively small number of insects pass the winter in the larval or active stage of the young. Of these, perhaps the best known is the brown "woolly worm" or "hedgehog caterpillar," as Larvæ of Insects in Winter. it is familiarly called. It is thickly covered with stiff black hairs on each end, and with reddish hairs on the middle of the body. These hairs appear to be evenly and closely shorn, so as to give the animal a velvety look; and as they have a certain degree of elasticity, and the caterpillar curls up at the slightest touch, it generally manages to slip away when taken into the hand. Beneath loose bark, boards, rails and stones, this caterpillar may be found in mid-winter, coiled up and apparently lifeless. On the first bright, sunny days of spring it may

Fig. 88—Hedgehog Caterpillar.

be seen crawling rapidly over the ground, seeking the earliest vegetation which will furnish it a literal "breakfast." In April or May the chrysalis, surrounded by a loose cocoon formed of the hairs of the body interwoven with coarse silk, may be found in situations similar to those in which the larva passed the winter. From this, the perfect insect, the Isabella tiger moth, *Pyrrharctia isabella* Smith, emerges about the last of June. It is a medium sized moth, dull orange in color, with three rows of small black spots on the body, and some scattered spots of the same color on the wings.

By breaking open rotten logs one can find in mid-winter the grubs or larvæ of many of the wood-boring beetles, and, beneath logs and stones near the margins of ponds and brooks, hordes of the maggots or larvæ of certain kinds of flies may often be found huddled together in great masses. The larvæ of a few butter-flies also live over winter beneath chips or bunches of leaves near the roots of their food plant, or in webs of their own construction, which are woven on the stems close to the buds, whose expanding leaves will furnish them their first meal in spring.

Many insects pass the winter in the quiescent or pupal stage; a state exceedingly well fitted for hiber-

nating, requiring, as it does, no food, and giving
plenty of time for the marvelous

Pupæ of Insects in Winter.

changes which are then undergone.
Some of these pupæ are enclosed in
dense silken cocoons, which are bound to the twigs of

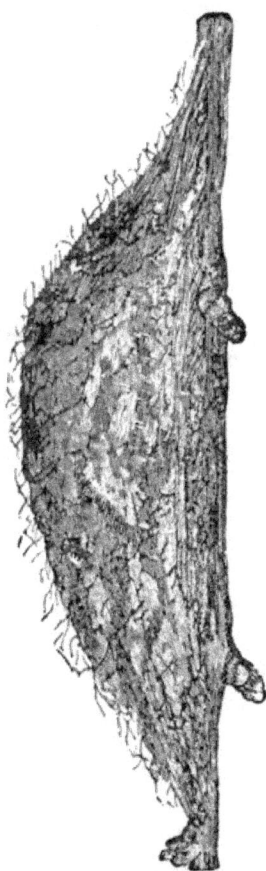

the plants upon which the larvæ feed, and thus they swing
securely in their silken ham-
mocks through all the storms
of winter. Perhaps the most
common of these is that of the
brown Cecropian moth, *Attacus
cecropia* L., the large oval cocoon of which is a conspicuous object in the winter on the
twigs of our common shade
and fruit trees. Many other
pupæ may be found beneath
logs or on the under side of
bark, and usually have the
chrysalis surrounded by a thin
covering of hairs, which are
rather loosely arranged. A
number pass the cold season
in the earth with no protective
covering whatever. Among
these is a large brown chrysalis
with a long tongue case bent
over so as to resemble the
handle of a jug. Every farm
boy has plowed or spaded it
up in the spring, and it is but

Fig. 89—Cocoon of Cecropian
Moth.

the pupa of a large sphinx moth, *Protoparce celeus*
Hub., the larva
of which is the
great green
worm, with a
"horn on its
tail," so com-

Fig. 90—Chrysalis of Tomato Worm.

mon on tomato plants in the late summer.

Each of the winter forms of insects above mentioned
can withstand long and severe cold weather—in fact,
may be frozen solid for weeks and retain life and
vigor, both of which are shown when warm weather
and food appear again. Indeed, it is not an unusually
cold winter, but one of successive thawings and freez-
ings, which is most destructive to insect life. A mild
winter encourages the growth of mold which attacks
the hibernating larvæ and pupæ as soon as, from
excess of rain or humidity, they become sickly; and
it also permits the continued activity of insectivorous
mammals and birds. Thus, moles, shrews, and field
mice, instead of burying themselves deeply in the
ground, run about freely during an open winter and
destroy enormous numbers of pupæ; while such birds
as the woodpeckers, titmice, and chickadees are constantly on the alert, and searching in every crevice
and cranny of fence and bark of tree for the hibernating larvæ.

Of the creeping, wingless creatures, which can ever
be found beneath rocks, rails, chunks, and especially
beneath those old decaying logs which are half buried
in the rich vegetable mold, the myriapods, or "thousand-legs," deserve more than a passing notice. They

21

are typical examples of that great branch of the animal kingdom known as *arthropods*, which comprises all insects and crustaceans. Each arthropod has the body composed of rings placed end to end and bearing jointed appendages, and in the myriapods each ring and its appendages can be plainly seen; whereas in the higher forms of the branch many of the rings are so combined as to be very difficult to distinguish.

Full forty kinds of myriapods occur in any area comprising one hundred square miles in the eastern United States. About twenty-five of them go by the general name of "thousand-legs" or millipedes, as each has from forty to fifty-five cylindrical rings in the body, and two pairs of legs to each ring. The other fifteen belong to the "centipede" group, the body consisting of about sixteen flattened segments, or rings, each bearing a single pair

Myriapods in Winter.

Fig. 91—Millipede or "Thousand Legs."

of legs. When disturbed, the "thousand-legs" generally coils up and remains motionless, shamming death, or "playing possum," as it is popularly put, as a means of defense; while the centipede scampers hurriedly away and endeavors to hide beneath leaf, chip, or other protecting object.

All those found in the Northern States are perfectly harmless, the true centipede, whose bite is reputed much more venomous than it really is, being found only in the South. True, some of the centipede group can pinch rather sharply with their beetle-like jaws;

and one, our largest and most common species, a brownish red fellow about three inches long and without eyes, can even draw blood if its jaws happen to strike a tender place. When handled, it always tries to bite, perhaps out of revenge for the abominably long Latin name given it by its describer. In fact the name is longer than the animal itself—*Sco-lo-po-cryp-tops sex-spi-no-sus* (Say)—being its cognomen in full. With such a handle attached to it, who can blame it for attempting to bite?

Fig. 92—Centipede.

Yet, to the scientist up on his Latin, each part of the above name bears a definite and tangible meaning. All the myriapods found in the woods and fields feed upon decaying vegetation, such as leaves, stems of weeds, and rotten wood, and in winter three or four species can usually be found within or beneath every decaying log or stump. One spe-

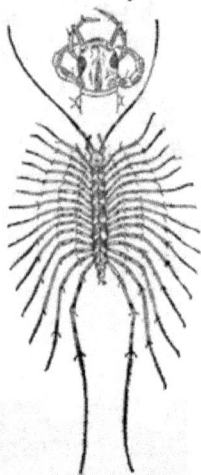

Fig. 93—Wall-sweeper.
(Two-thirds natural size; also head much enlarged. After Lintner.)

cies with very long legs, *Scutigera forceps* (Raf.), is often found in damp houses or in cellars. It is sometimes called the "wall-sweeper," on account of its rapid ungainly gait, and is even reputed to prey upon cockroaches and other household pests.

Spiders, which do not undergo such changes as do most of the common, six-footed insects, winter either

Spiders in Winter.

as eggs or in the mature form. The members of the "sedentary" or web-spinning group, as a rule, form nests in late autumn, in each of which are deposited from fifty to eighty eggs, which survive the winter and hatch in the spring, as soon as the food supply of gnats, flies

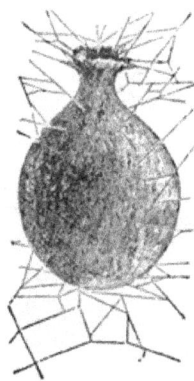

Fig. 94—Balloon-shaped Nest of Spider. (After Comstock.)

and mosquitoes appear. The different forms of spiders' nests are very interesting objects of study. Some are those close-spun, flat, button-shaped objects, about half an inch in diameter, which are so common in winter on the under side of bark, chunks and flat rocks. Others are balloon-shaped and attached to weeds. Within the latter the young spiders often hatch in early winter, make their first meal off their empty egg cases, and then begin a struggle for existence, the stronger preying upon the weaker until the south winds blow again, when they emerge and scatter far and wide in search of more nutritious sustenance.

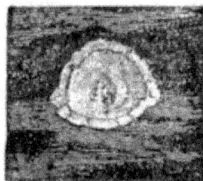

Fig. 95—Spider's Nest. (After Comstock.)

The "wandering" spiders never spin webs, but run actively about and pounce upon their prey with a tiger-like spring. Six or eight of the larger species of this group winter in the mature form beneath logs and chunks, being often frozen solid during cold weather, but thawing out as healthy as ever when the temperature rises. Retiring beneath the loose-fitting bark of hickory or maple trees, a number of the smaller tube-weaving spiders construct about themselves a protecting web of many layers of the finest silk. Within this snug retreat they lie from November until April—a handsome, small, black fellow, with green jaws and two orange spots on his abdomen, being the most common species found motionless within this seeming shroud of silk on a day in mid-winter.

In any Northern State as many as four hundred* different kinds of the six-footed or true insects, in the winged or adult stage, may be taken in winter by any one who is so disposed, and knows where to search for them. Among the *Orthoptera*, the "grouse grass-hoppers" live during the cold season beneath the loose bark of logs, or beneath the bottom rails of the old Virginia worm fences. From these retreats every warm, sunny day tempts them forth in numbers. On such occasions the earth seems to swarm with them, as they leap before the intruder, their hard bodies striking the dead leaves with a sound similar to that produced by falling hail. The common field cricket belongs also to the *Orthoptera*, and the young of various sizes winter under rails and logs, bidding defiance to Jack Frost from within a little burrow or pit beneath the protecting shelter.

---

*See Psyche, 1895 and 1896, for notes on 286 species of *Coleoptera*, 64 of *Hemiptera* and 18 of *Orthoptera* taken by the writer in Vigo county, Indiana, during the winter months.

The true bugs, or *Hemiptera*, hibernate in similar places; squash bugs, chinch bugs, "stink" bugs, and others being easily found in numbers beneath loose bark or hidden between the root leaves of mullein and other plants.

Nearly three hundred species of *Coleoptera*, or beetles, occupy similar positions. Almost any rotten log or stump when broken open discloses a half dozen or more "horn"

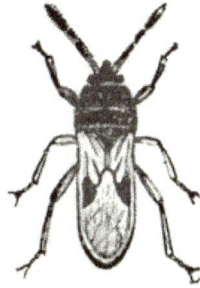

Fig. 96—Chinch Bug.

(Enlarged five times.)

Beetles in Winter.

or "bess beetles," *Passalus cornutus* L., great, shining, clumsy, black fellows with a curved horn on the head. They are often

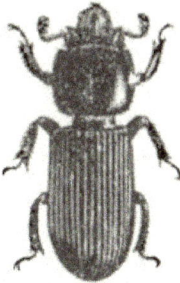

Fig. 97—Horn or Bess Beetle.

utilized as horses by country children, the horn furnishing an inviting projection to which may be fastened, by a thread or cord, chips and pieces of bark to be dragged about by the strong and never lagging beast of burden. When tired of "playing horse" they can make of the insect an instrument of music; for, when held by the body, it emits a creaking, hissing noise, produced by rubbing the abdomen up and down against the inside of the hard, horny wing covers. This beetle passes its entire life in cavities in the rotten wood on which it feeds, and when it wishes a larger or more commodious home it has only to eat the more.

The handsome and beneficial lady beetles winter beneath fallen leaves or between and beneath the root leaves of the mullein and the thistle. Our most common species, the thirteen-spotted lady beetle, *Megilla maculata* De G., is gregarious, collecting together by thousands on the approach of cold weather, and lying huddled up like sheep until a breath of spring gives them the signal to disperse. Snout beetles galore can be found beneath piles of weeds near streams and the borders of ponds or beneath chunks and logs in sandy places. All are injurious, and the farmer by burning their hibernating places in winter can cause their destruction in numbers. Rove beetles, ground beetles, and many others live deep down in the vegetable mold beneath old logs, where they are, no doubt, as secure from the breath of the ice king as if they had followed the swallow to the tropics.

Fig. 98—Thirteen-spotted Lady Beetle.

Of the *Diptera*, or flies, but few forms winter in the perfect state, yet the myriads of house flies and their kin, which next summer will distract the busy housewife, are represented in winter by a few isolated individuals which creep forth occasionally from crevice or cranny and greet us with a friendly buzz.

Flies and Gnats in Winter.

In mid-winter one may also see in the air swarms of small, gnat-like insects. They belong to this order and live beneath the bark of freshly fallen beech and other logs. On warm, sunny days they go forth in numbers for a sort of rhythmical courtship; their movements while in the air being peculiar in that they usually rise and fall in the same vertical line—performing a curious aërial dance which is long continued.

Among the dozen or more butterflies and moths which winter in the perfect state, the most common and the most handsome is the "Camberwell beauty" or "mourning cloak," *Vanessa antiopa* L., a large butterfly whose wings are a rich purplish brown above, duller beneath and broadly margined with a yellowish band. It is often found in winter beneath chunks which are raised a short distance above the ground, or in the crevices of old snags and fence rails. It is then apparently lifeless, with the antennæ resting close along the back, above which the wings are folded. But one or two warm days are necessary to restore it to activity, and I have seen it on the wing as early as the 2d of March, hovering over the open flowers of the little snow trillium.

All the species of ants survive the winter as mature

Fig. 99—A Queen Bumble-bee.

forms, either in their nests in the ground or in huddled groups in half rotten logs and stumps; while here and there beneath logs a solitary queen bumble-bee, bald hornet, or yellow jacket is found —the sole representatives of their races.

Thus insects survive the winter in many ways and in many places, some as eggs, others as larvæ, still

others as pupæ, and a large number as adults—all
being able to withstand severe cold and yet retain
vitality sufficient to recover, live, grow, and replenish
the earth with their progeny when the halcyon days
of spring appear once more.

In the scale of animal life the vertebrates or back-
boned animals succeed the insects. Beginning with
the fishes, we find that in late autumn they mostly
seek some deep pool in pond or stream at the bottom
of which the water does not freeze.

Fishes in Winter. Here the herbivorous forms eke out a
precarious existence by feeding upon
the innumerable diatoms and other small plants which
are always to be found in water, while the carnivorous prey upon the herbivorous, and so maintain the
struggle for existence. The moving to these deeper
channels and pools in autumn and the scattering in
the spring of the assembly which has gathered there
constitute the so-called "migration of fishes," which
is far from being so extensive and methodical as that
practiced by the migratory birds.

Many of the smaller species of fishes, upon leaving
these winter resorts, ascend small, clear brooks in
large numbers for the purpose of depositing their
eggs ; as, when hatched in such a place, the young will
be comparatively free from the attacks of the larger
carnivorous forms. Among the lowest vertebrates
often found in numbers in early spring in these
meadow rills and brooks is the lamprey, *Ammocœtes
branchialis* (L.), or "lamper eel," as it is sometimes
called. It has a slender, eel-like body, of a uniform
leaden or blackish color, and with seven purse-shaped

gill openings on each side. The mouth is fitted for sucking rather than biting, and with it they attach themselves to the bodies of fishes and feed on their flesh, which they scrape off with their rasp-like teeth. Later in the season they disappear from these smaller streams, probably returning in mid-summer to deeper water. Thoreau, who studied their habits closely, says of them: "They are rarely seen on their way down stream, and it is thought by fishermen that they never return, but waste away and die, clinging to rocks and stumps of trees for an indefinite period; a tragic feature to the scenery of the river bottoms worthy to be remembered with Shakespeare's description of the sea floor."

A few of the fishes, as the mud minnow and smaller catfishes, together with most frogs, turtles, and sala-

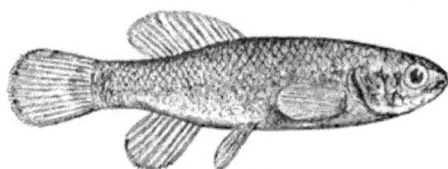

Fig. 100—Mud Minnow.
Umbra limi (Kirtland).

manders, on the approach of winter, burrow into the mud at the bottom of the streams and ponds, or beneath logs near their margins. There they live without moving about and with all the vital processes in a partially dormant condition, thus needing little if any food.

The box tortoise or "dry land terrapin," the common toad, and some salamanders burrow into the dry earth, usually going deep enough to escape frost; while snakes seek some crevice in the rocks or hole in the ground where they coil themselves together, oftentimes in vast numbers, and prepare for their winter's sleep. In an open winter this hibernation is often interrupted, the animal emerging from its retreat and seeking its usual summer haunts as though spring had come again. Thus I have, on one occasion, seen a soft-shelled turtle moving gracefully over the bottom of a stream on a day in late December, and have in mid-January captured snakes and salamanders from beneath a pile of driftwood, where they had taken temporary refuge.

With frogs, especially, this hibernation is not a perfect one, and there is a doubt if in a mild winter some species hibernate at all. For example, the little cricket frog or "peeper" has been seen many times in mid-winter alongside the banks of flowing streams, and during the open winter of 1888–89 numerous specimens of leopard and green frogs were seen on different occasions in December and January, while on February 18th they, together with the "peepers," were in full chorus.

Of our mammals, a few of the rodents or gnawers, as the ground-hogs, gophers and chipmunks, hibernate in burrows deep enough to escape the cold, and either feed on a stored supply of food, or, like the snakes and crayfish, do not feed at all.

Others, as the rabbits, field mice, and squirrels, are more or less active and forage freely on whatever they can find, eating many things which in summer they would spurn with scorn. To this class **The Muskrat in Winter.** belongs that intelligent but injurious animal the musquash or muskrat. Those which inhabit the rivers and larger streams live

in burrows dug deep beneath the banks, but those inhabiting sluggish streams and ponds usually construct
a conical winter house about three feet in diameter and
from two to three feet in height. These houses are
made of coarse grasses, rushes, branches of shrubs,
and small pieces of driftwood, closely cemented together with stiff, clayey mud. The top of the house
usually projects two feet or more above the water,
and when sun-dried is so strong as to easily sustain
the weight of a man. The walls are generally about
six inches in thickness and are very difficult to pull
to pieces. Within is a single circular chamber with a
shelf or floor of mud, sticks, leaves and grass, ingen-
iously supported on coarse sticks stuck endwise into
the mud after the manner of piles. In the center of
this floor is an opening, from which six or eight diverging paths lead to the open water without, so that
the little artisan has many avenues of escape in case
of danger. These houses are often repaired and used
for several winters in succession, but are vacated on
the approach of spring. During the summer the
muskrat is, in the main, a herbivorous animal, but in
winter necessity develops its carnivorous propensities
and it feeds then mainly upon the mussels and crayfish which it can dig from the bottom of the pond or
stream in which its house is built.

The bats pass the winter in caves, the attics of
houses and barns, or in hollow trees, hanging downward by their hind claws. Motionless for months
they thus remain, and those in the more exposed situations are, doubtless, frozen solid. Yet, in time, their
blood flows freely once again and they become as

XV

WINTER HOUSE OF MUSKRAT.

expert on the wing as though the year were one continual jubilee of insect chasing, and frost and snow were to them unknown.

All the carnivora, or flesh-eaters, as the mink, skunk, opossum, fox and wolf, are in winter active and voracious, needing much food to supply the necessary animal heat of the body. Hence they are then much more bold than in summer, and the hen yard or sheep pen of the farmer is too frequently called upon to supply this extra demand.

But of all our animals it seems to me the birds have solved the winter problem best. Possessing an enduring power of flight and a knowledge of a southern sunny sky, beneath which food is plentiful, they alone of all the living forms about us have little fear of the coming of the frost. True, fifty or more species remain in each of the Northern States during the cold season, but they are hardy birds which feed mainly upon seeds, as the snow-bird and song sparrow; on flesh, as the hawks and crows; or on burrowing insects, as the nuthatches and woodpeckers.

Such are some of the solutions to the problem of life in winter which the plants and animals about us have worked out; such some of the forms which they undergo; the places which they inhabit.

To the thinking mind a knowledge of these solutions but begets other and greater problems, such as how can a living thing be frozen solid for weeks and yet retain vitality enough to fully recover? How can a warm-blooded animal sleep for months without partaking of food or drink? And, greater than either, what is that which we call life?

I hold in my hand two objects, similar in size, color, organs, everything—twins from the same mother in all outward respects. One pulsates and throbs with that which we call "life." It possesses heat, bodily motion, animal power. The other is cold, motionless, pulseless, throbless—a thing of clay. What is that "life" which the one possesses and the other lacks? Ah, there's the rub! With the wisest of men we can only answer, "*Quien sabe?*" (Who knows?)

A SEEKER AFTER GOLD.

A few months ago I spent a day in the wildest part of Brown county, looking for traces of that precious metal which Dame Rumor says exists in quantity in the fine sand and silt along the streams and in the valleys thereabouts. My eyes, unaccustomed to such work, saw no traces of yellow amidst the gray and the black, and I was beginning to doubt its existence except in the mind of some enthusiastic seeker, when, on turning a sharp bend in a stream, I came suddenly upon an old man, weather-beaten, roughly clad, gaunt of figure and haggard of face, who was bending over a pan of moist sand and silt which he was shaking with a rocking motion to and fro. So busily engaged in his work was he that he did not notice my approach, and I stood beside him and heard his ejaculation of delight as he reached down and picked from the bottom of the pan a piece of gold about double the size of a grain of wheat. "Ah, my little beauty, I have found you at last," said he; then for the first time noticing my presence, he sprang to his feet with an exclamation of surprise, letting fall his pan in his excitement. His locks were unkempt, his face begrimed, but his eyes sparkled with more than ordinary brilliancy and through them was revealed the soul of a man who was an enthusiast in his chosen work—the search for gold.

I engaged him in conversation and found that he had been washing the sands for three weeks and longer, but with very poor success—less than an ounce being the total result of his labor—all in small grains, the one just found being the largest. "I am doing this work more for pleasure than for profit," said he. "I do not have to work, for I spent many years prospecting in the west and finally found a paying lead, sold out and came home, not rich, but with enough to keep me from want for the remainder of my days. The bright Indian summer weather of the past few weeks has tempted me forth and again have I been seeking the yellow grains in the sands and gravel of these streams."

Talking farther with him, I found him to be a man of fine education—a graduate of an eastern college, but a life long rover—who, like thousands of others, had given his years, more than thirty of them, to the search for gold—forsaking friends, society, all, in a vain seeking for great wealth.

He invited me to his tent, pitched on a near-by sunny slope, and there for an hour or longer entertained me with anecdotes of his prospecting life among the hills and mountains of the distant west. Noticing a number of books in the tent, I led him to talk of them, and found his knowledge of poetry to be extensive, Bryant and Wordsworth being his acknowledged favorites. As I was leaving he took from between the covers of one of the volumes a folded piece of paper and handing it to me said : "At times I indulge a little in poetry myself. Here is a copy of my latest verse. When you get home, read it, and if

.

we ever meet again tell me what you think of it." I took it, and, having heard nothing from him since, I venture to reproduce it here, as it shows that he realized something of what he had lost in the years gone by, and also that in poetry as in life his favorite theme was gold. It was inscribed "To Some Nuggets of Brown County Gold," and ran as follows:

TO SOME NUGGETS OF BROWN COUNTY GOLD.

Gold, gold,
Tiny nuggets of yellow gold,
Brought from the north by a glacier cold,
Borne with the sands and the pebbles old
To the vales of Brown, and there out-doled
To remain alway.

Gold, gold,
Ever a curse to man the bold,
Luring him forth from his father's fold
To lands far away, where hills uprolled
Forever will be; and bells untolled
Till the judgment day.

Gold, gold,
How many faces are pinched and old,
How many hearts once warm and bold
To-day are timid, and sere and cold,
How many bodies beneath the mold,
For search of thee?

Gold, gold,
To mortal ear the half's not told
Of men's souls lost and women's sold
For sake of such baubles as these I hold.
Then, curses upon thee, yellow gold,
Forever be.

INDEX.

www.ingramcontent.com/pod-product-compliance
Lightning Source LLC
Chambersburg PA
CBHW021358210326

41599CB00011B/921